THE SCIENCE WE LIVE BY

Nibedita Sahu

The Science We Live

(Understanding Our Everyday World)

by **Nibedita Sahu**

Written by **Nibedita Sahu**

Preface:

As I began the journey of writing this book, I was amazed by the complex web of science woven into the fabric of our everyday lives. From the moment we wake up until we retire to bed, we are surrounded by the images of scientific principles, often without even realizing it. "The Science We Live By" seeks to illuminate these hidden connections and disclose the beauty and complexity of the scientific world that forms our existence. Whether you're a seasoned scientist or simply curious about the forces at play in your daily routines, I invite you to start on this enlightening exploration of the science we live by.

Table of Contents:

Author's Note:

Dear Readers,

As you begin on this exploration of "The Science We Live By," I want to extend a warm welcome and share a bit about myself. My name is Nibedita, and I am deeply passionate about mathematics and the wonders of science. With a background in mathematics, I have always been drawn to the beauty of logical reasoning and problem-solving.

I am someone who thrives on curiosity, constantly seeking to understand the world around me and uncovering insights clearly from every day. Observing the intricacies of everyday life has been a source of inspiration for me, fueling my desire to delve deeper into the underlying science that rules our existence.

With a professional belief in the power of logic and science, I approach life's decisions and challenges with a rational mindset, guided by evidence and reason. My path has further expanded my appreciation for the role of technology in shaping our world, and I am always eager to explore new advancements and acquire new skills.

It's my hope that through "The Science We Live By," you will gain a newfound appreciation for the wonders of science that surround us and perhaps discover your own passion for exploration and discovery.

Warm regards,

Nibedita Sahu

Detailed Description about the book:

In "**The Science We Live By**: Understanding Our Everyday World," author Nibedita takes readers on a fascinating path through complex shadows of scientific principles that sustain our daily lives. With a keen eye for observation and a passion for problem-solving honed through a background in mathematics, she tells the hidden science behind the seemingly ordinary occurrences that shape our existence.

From the mathematical precision of timekeeping to the physics of light and sound, each chapter of this illuminating book delves into a different aspect of our world, revealing the fascinating interplay between science and our everyday experiences. Through engaging narratives and insightful explanations, readers will gain a deeper understanding of the role that mathematics, physics, information technology, chemistry and biology play in shaping our reality.

Nibedita's exploration extends beyond mere explanation, offering readers the opportunity to uncover the beauty of science in their own lives. Whether unraveling the mysteries of probability and statistics or exploring the wonders of artificial intelligence and machine learning, "The Science We Live By" invites readers to become active participants in the discovery of the world around them.

With a blend of intellectual difficulty and accessible prose, Nibedita explains complex scientific concepts, making them accessible to readers of all backgrounds. Whether you're a professional or simply curious about the forces at play in your daily routines, this book promises to enlighten and inspire, offering a newfound appreciation for the science that surrounds us all.

Prepare to begin on a journey of discovery, where every page reveals new insights and every chapter illuminates the profound impact of science on our everyday world. "The Science We Live By" is not just a book — it's an invitation to see the world through the lens of science and unlock the mysteries of the universe that await us all.

Book Outline:

Chapter 1: The Mathematics of Time

1.1. Exploring the concept of time through different mathematical models

1.2. The role of mathematics in measuring and predicting time

Chapter 2: The Physics of Light

2.1. Understanding the nature of light and its everyday applications

2.2. The role of physics in explaining phenomena like rainbows and mirages

Chapter 3: The IT Behind Our Digital World

3.1. Exploring how information technology shapes our daily lives

3.2. The science behind the internet, social media, and digital communication

Chapter 4: The Mathematics of Chance

4.1. Understanding probability and its role in everyday events

4.2. The use of statistics in interpreting data and making predictions

Chapter 5: The Physics of Motion

5.1. Exploring the principles of motion in everyday activities like sports and transportation

5.2. The role of physics in explaining why things move the way they do

Chapter 6: The IT in Everyday Devices

6.1. Understanding how IT powers common devices like smartphones and home appliances

6.2. The science behind computer hardware and software

Chapter 7: The Mathematics of Nature

7.1. Exploring patterns and sequences in the natural world

7.2. The role of mathematics in understanding phenomena like fractals and the Fibonacci sequence

Chapter 8: The Physics of Sound

8.1. Understanding the nature of sound and its role in communication and music

8.2. The science behind acoustics and noise cancellation

Chapter 9: The IT Behind Artificial Intelligence

9.1. Exploring the principles of artificial intelligence and machine learning

9.2. The role of IT in powering modern AI applications

Chapter 10: The Chemistry of Life

10.1. Exploring the role of chemistry in everyday life

10.2. The science behind common chemical reactions and substances

Chapter 11: The Zoology of Our World

11.1. Understanding the diversity of animal life and its impact on our daily lives

11.2. The science behind animal behavior and ecosystems

Chapter 12: The Botany of Our Surroundings

12.1. Exploring the world of plants and their role in our everyday lives

12.2. The science behind plant growth, photosynthesis, and plant-based products

Chapter 13: The Mathematics of Finance

13.1. Understanding financial concepts like interest, investment, and risk through mathematics

13.2. The role of maths in personal finance and economic forecasting

Chapter 14: The Physics of Energy

14.1. Exploring the principles of energy conservation and transformation

14.2. The role of physics in understanding renewable energy sources and climate change

Chapter 15: The IT of the Future

15.1. Predicting future trends in information technology

15.2. The potential impact of quantum computing, virtual reality, and other emerging technologies

The Science We Live By

(Understanding Our Everyday World)

Nibedita Sahu

Chapter 1: The Mathematics of Time

Time is a fundamental aspect of our existence, a dimension that underpins every moment and action. The mathematics of time allows us to measure, understand, and quantify this enigmatic concept. By examining time mathematically, we can appreciate its structure and its impact on our daily lives, from the ticking of a clock to the movements of celestial bodies.

In this chapter, we will explore the various ways in which time is measured and represented mathematically. We begin with the basics of units of time, such as seconds, minutes, and hours, and how they form the foundation of our perception of time. We will also delve into more advanced topics such as time dilation, which occurs when the relative speed of objects affects the passage of time, as described by Einstein's theory of relativity.

Throughout the chapter, we will consider the tools and methods used to track and measure time accurately, from ancient sundials to modern atomic clocks. We will also explore the relationship between time and space, known as spacetime, and how it shapes our understanding of the universe.

Finally, we will discuss the role of time in various aspects of our daily lives, including scheduling, productivity, and the natural cycles of sleep and wakefulness. By understanding the mathematics of time, we gain insight into how to make the most of each precious moment.

1.1. Exploring the concept of time through different mathematical models

Time is a complex concept that can be explored and understood using various mathematical models. These models provide different perspectives on the nature and behavior of time, allowing us to understand how it affects our daily lives and the universe at large.

1.1.1. Linear Time:

The most familiar model of time is the linear model, which views time as a straight line that progresses in one direction—from the past through the present and into the future. In this model, time can be quantified and measured in consistent units such as seconds, minutes, and hours. This linear progression is the foundation for calendars and clocks, enabling us to plan and schedule activities based on the steady, unidirectional flow of time.

1.1.2. Cyclical Time:

Another model views time as cyclical, focusing on the repetition of natural phenomena such as seasons, lunar phases, and circadian rhythms. In this perspective, time is not a linear progression but rather a continuous loop where events repeat in regular patterns. This model aligns with many ancient cultures and philosophies that emphasize the recurring cycles of nature.

1.1.3. Relativistic Time:

Albert Einstein's theory of relativity introduced a groundbreaking model of time that accounts for the effects of speed and gravity on the passage of time. According to this model, time can dilate or contract depending on an object's relative speed or the strength of a gravitational field. This phenomenon is evident in experiments with atomic clocks on airplanes and in satellites, where clocks tick slightly slower or faster due to differences in velocity and gravitational forces.

1.1.4. Discrete Time:

In some mathematical models, time is treated as a discrete quantity rather than a continuous one. This means that time advances in distinct, separate intervals rather than a smooth, unbroken flow. This model is often used in computer science and digital systems where processes are executed in discrete time steps, providing a framework for analyzing and designing complex algorithms and simulations.

1.1.5. Multidimensional Time:

A more abstract approach considers the possibility of multidimensional time, where time may exist in more than one dimension. This model is still largely theoretical and not yet widely accepted, but it raises intriguing questions about the nature of time and its relationship with space.

By exploring these different mathematical models of time, we gain a richer understanding of how time influences our world and how we can measure, interpret, and interact with it in various contexts. Each model offers unique insights and applications, contributing to our overall comprehension of the concept of time.

1.2. The role of mathematics in measuring and predicting time

Mathematics plays a crucial role in measuring and predicting time, providing the tools and frameworks necessary to understand and navigate the complex dynamics of temporal phenomena. From the development of precise timekeeping instruments to the prediction of future events, mathematical concepts and techniques are foundational to our ability to measure, represent, and forecast time.

1.2.1. Units of Time and Conversion:

Mathematics allows us to standardize units of time such as seconds, minutes, hours, and days, enabling consistent measurement and comparison across different systems and contexts. By using mathematical conversions, we can seamlessly switch between various time units for different applications, from everyday scheduling to scientific experiments.

1.2.2. Timekeeping and Accuracy:

Mathematics is essential in the design and calibration of timekeeping devices such as clocks and watches. By incorporating mathematical principles, these instruments can maintain high levels of accuracy, ensuring reliable and precise timekeeping. Modern atomic clocks, for instance, rely on quantum mechanics and mathematical algorithms to measure time based on the vibrations of atoms, providing unparalleled precision.

1.2.3. Predicting Celestial Events:

Mathematical models are used to predict celestial events such as eclipses, planetary transits, and lunar phases. These models take into account the movements and positions of celestial bodies in their orbits, using mathematical equations to calculate when specific events will occur. This ability to predict celestial events has practical applications in fields such as astronomy, navigation, and agriculture.

1.2.4. Scheduling and Planning:

Mathematics aids in organizing time efficiently through scheduling and planning. By using mathematical concepts such as intervals, sequences, and optimization, we can create schedules that maximize productivity and minimize conflicts. Project management techniques such as the critical path method (CPM) and Gantt charts rely on mathematical principles to organize tasks and timelines.

1.2.5. Time Series Analysis:

Mathematics is instrumental in the analysis of time series data, which involves studying patterns and trends over time. Techniques such as moving averages, exponential smoothing, and autoregressive models help identify and predict future trends in various fields, including economics, finance, weather forecasting, and population studies.

1.2.6. Relativity and Time Dilation:

The theory of relativity, which includes time dilation, relies on mathematical equations to describe the relationship between time, space, velocity, and gravity. By applying these equations, scientists can predict how time behaves differently depending on an object's speed and proximity to massive objects.

In summary, mathematics provides the foundation for measuring and predicting time, from the creation of precise timekeeping instruments to the forecasting of future events. Through mathematical models and techniques, we gain a deeper understanding of how time influences our world and how we can navigate it effectively.

Chapter 2: The Physics of Light

Light is a fundamental aspect of our universe, playing a crucial role in our perception of the world and our ability to navigate through it. It is the primary medium through which we see and understand our surroundings. The study of light, or optics, encompasses a wide range of phenomena and applications, from how light interacts with matter to the behavior of light in different environments.

In this chapter, we will explore the physics of light, examining its nature, properties, and the ways in which it influences our lives. We begin with the basics of light, including its wave-particle duality, frequency, wavelength, and speed. Understanding these concepts allows us to comprehend how light travels and interacts with different materials.

Next, we will delve into the different behaviors of light, such as reflection, refraction, and diffraction, and how these phenomena give rise to everyday occurrences like rainbows and mirages. We will also explore the electromagnetic spectrum, which includes not only visible light but also a range of other wavelengths such as ultraviolet, infrared, and radio waves.

Throughout the chapter, we will examine how the study of light has led to technological advances, such as the development of lasers, fiber optics, and imaging technologies. These innovations have revolutionized fields such as communication, medicine, and entertainment.

Finally, we will discuss the profound impact of light on our daily lives, from providing illumination to enabling scientific discoveries and artistic expression. By understanding the physics of light, we gain insight into one of the most fundamental and versatile aspects of our world.

2.1. Understanding the nature of light and its everyday applications

Light is a form of electromagnetic radiation that is essential to our ability to perceive and interact with the world around us. It exhibits both wave-like and particle-like properties, a phenomenon known as wave-particle duality. This dual nature allows light to exhibit behaviors such as interference, diffraction, and the photoelectric effect.

2.1.1. The Nature of Light:

Light travels in electromagnetic waves, which consist of oscillating electric and magnetic fields. The frequency and wavelength of these waves determine the characteristics of the light, such as its color and energy. Visible light, the portion of the electromagnetic spectrum that we can see, ranges from approximately 400 to 700 nanometers in wavelength.

Light also behaves as a stream of particles called photons, which carry energy and momentum. This particle-like nature of light becomes evident in phenomena such as the photoelectric effect, where light can eject electrons from a material, generating an electric current.

2.1.2. Everyday Applications of Light:

Light has numerous everyday applications that have become indispensable in modern society:

- **Illumination:** Artificial lighting is a fundamental application of light, providing visibility in homes, workplaces, and public spaces. Technologies such as LED and fluorescent lighting offer energy-efficient options for illumination.

- **Communication:** Light is used in fiber optic cables to transmit data at high speeds over long distances. This technology forms the backbone of modern internet and telecommunication networks.
- **Medicine:** Light-based technologies play a significant role in medical diagnostics and treatment. For instance, endoscopes use light to visualize internal organs, while laser therapies are used to treat various medical conditions.
- **Photography and Imaging:** Cameras and imaging devices rely on light to capture visual information. Advances in sensor technology have improved image quality and enabled applications such as facial recognition and satellite imaging.
- **Entertainment:** Light is essential in the entertainment industry, from stage lighting in theaters to video displays in televisions and projectors.
- **Energy Production:** Solar panels harness sunlight to generate electricity, providing a renewable and sustainable energy source.
- **Astronomy:** Light from celestial bodies allows astronomers to study the universe. Observing different wavelengths of light, such as infrared and X-ray, provides insights into the composition and behavior of distant stars and galaxies.

The versatility and ubiquity of light make it a critical element of modern life. Understanding its nature and how it can be manipulated allows us to continue developing new technologies and applications that benefit society.

2.2. The role of physics in explaining phenomena like rainbows and mirages

Physics plays a key role in explaining natural optical phenomena such as rainbows and mirages, which result from the interaction of light with different materials and conditions in the environment. These phenomena illustrate the unique behaviors of light, including reflection, refraction, and dispersion, and offer us insight into the intricate interplay between light and the world around us.

2.2.1. Rainbows:

Rainbows are one of the most captivating optical phenomena, and their formation can be explained by the physics of light interacting with water droplets in the atmosphere.

- **Refraction:** When sunlight enters a water droplet, the light changes direction due to refraction, bending as it passes from air into the denser water medium.
- **Dispersion:** Once inside the droplet, the light is dispersed into its component colors due to varying refraction angles for different wavelengths (colors) of light. This separation of colors creates the familiar spectrum of the rainbow.
- **Reflection:** The light is then internally reflected off the inner surface of the droplet, allowing it to change direction once again.
- **Refraction and Emergence:** Finally, the light exits the droplet, undergoing another refraction as it leaves the droplet and re-enters the air. The bending and separation of colors are more pronounced at this stage, contributing to the vividness of the rainbow.

The combination of these processes leads to the appearance of a circular arc of colors in the sky when the viewer stands with the sun behind them and rain in front.

2.2.2. Mirages:

Mirages are optical illusions that occur when light bends as it travels through layers of air with different temperatures and densities.

- **Refraction and Layered Air:** In a mirage, light passes through layers of air with varying temperatures, which causes changes in air density. Light travels faster through warmer, less dense air and slower through cooler, denser air.
- **Inferior Mirage:** This type of mirage occurs when cooler air is above and warmer air is below. Light rays traveling from the sky downwards bend as they encounter the temperature gradient, making it appear as though a reflective surface (e.g., water) exists on the ground.
- **Superior Mirage:** This type of mirage happens when warmer air is above and cooler air is below. Light from objects below the horizon can be bent over the curve of the Earth and into the viewer's line of sight, making distant objects appear distorted, elevated, or stretched.

Mirages are common in deserts and over bodies of water where significant temperature gradients can occur. They create illusions of water, floating objects, or distorted landscapes.

Both rainbows and mirages offer beautiful examples of how the physics of light can create mesmerizing visual effects in the natural world. Understanding the scientific principles behind these phenomena not only enhances our appreciation for their beauty but also provides valuable insights into the behavior of light and its interactions with the environment.

Chapter 3: The IT Behind Our Digital World

The digital world has transformed every aspect of modern life, from the way we communicate and work to how we access information and entertain ourselves. This transformation is made possible by information technology (IT), which encompasses a wide range of hardware, software, and systems that facilitate the creation, processing, storage, and transmission of digital data.

In this chapter, we will explore the IT behind our digital world, examining the technologies, concepts, and infrastructure that enable our interconnected society. We begin by delving into the fundamentals of computing, including binary code, processors, and memory, which serve as the foundation of digital devices.

Next, we will explore the world of software, which includes operating systems, applications, and programming languages. Software is the bridge between hardware and users, providing the functionality and tools needed to perform a variety of tasks.

Communication networks play a crucial role in our digital world, enabling data transmission across the globe. We will discuss the structure and workings of the internet, as well as the technologies that make high-speed data transfer and connectivity possible.

Throughout the chapter, we will touch on topics such as cloud computing, data security, and artificial intelligence, all of which are essential components of modern IT. These technologies offer new opportunities and challenges, shaping the way we interact with digital devices and information.

Finally, we will consider the impact of IT on society, including issues such as digital privacy, cybersecurity, and the digital divide. As we become increasingly reliant on digital technology, understanding the IT behind our digital world is essential for making informed decisions and navigating the complexities of the digital age.

3.1. Exploring how information technology shapes our daily lives

Information technology (IT) has become an integral part of our daily lives, influencing how we communicate, work, learn, shop, and entertain ourselves. Its impact can be seen in almost every aspect of modern life, enabling convenience, efficiency, and connectivity on an unprecedented scale. Here are some key ways in which IT shapes our daily lives:

3.1.1. Communication:

- **Instant Messaging and Video Calls:** IT allows us to communicate instantly with people around the world through messaging apps and video conferencing tools. This seamless connectivity has revolutionized both personal and professional communication.
- **Social Media:** Social networking platforms enable us to share experiences, stay connected with friends and family, and access news and information in real-time. These platforms have transformed how we engage with the world and each other.

3.1.2. Work and Productivity:

- **Remote Work:** IT has made it possible for many people to work remotely from anywhere with an internet connection. This flexibility has become increasingly important, especially during global events such as the COVID-19 pandemic.
- **Collaboration Tools:** Cloud-based productivity tools, project management software, and online collaboration platforms enable teams to work together efficiently, regardless of their physical location.

3.1.3. Access to Information and Education:

- **Online Learning:** IT has democratized access to education through online courses, webinars, and e-learning platforms. This allows people to gain new skills and knowledge from anywhere in the world.
- **Search Engines and Digital Libraries:** Search engines provide quick access to vast amounts of information on virtually any topic. Digital libraries offer access to books, journals, and other resources for research and learning.

3.1.4. Entertainment:

- **Streaming Services:** IT enables on-demand streaming of movies, TV shows, music, and podcasts. This shift away from traditional media consumption provides greater choice and convenience for consumers.
- **Gaming:** Video games, both on consoles and mobile devices, offer immersive experiences and multiplayer online platforms that connect players worldwide.

3.1.5. E-commerce and Finance:

- **Online Shopping:** IT has revolutionized retail through e-commerce platforms, enabling consumers to shop for a wide variety of goods and services from the comfort of their homes.
- **Digital Banking and Payments:** Online banking and mobile payment apps make managing finances and conducting transactions easier and more accessible.

3.1.6. Healthcare and Wellness:

- **Telehealth:** IT facilitates virtual medical consultations and remote monitoring, improving access to healthcare for people in remote or underserved areas.
- **Fitness and Wellness Apps:** Mobile apps and wearables help individuals track their health and fitness goals, providing personalized recommendations and feedback.

3.1.7. Navigation and Travel:

- **GPS and Mapping Services:** IT offers real-time navigation and mapping services that guide us on the road, helping us plan routes and avoid traffic.

In summary, information technology has become an indispensable part of our daily lives, enhancing convenience, connectivity, and access to information and services. As IT continues to evolve, its impact on our lives will only deepen, opening up new possibilities and challenges in the process.

3.2. The science behind the internet, social media, and digital communication

The internet, social media, and digital communication have transformed the way we connect with each other and access information. These technologies rely on a foundation of scientific principles and engineering innovations that enable data to be transmitted quickly and efficiently across the globe.

3.2.1. The Internet:

- **Packet Switching:** The internet operates on the principle of packet switching, where data is divided into small packets that travel independently through the network. These packets are routed based on destination addresses and reassembled at the receiving end to form the original message or data.
- **IP Addresses and Protocols:** Every device on the internet is assigned a unique IP (Internet Protocol) address, which serves as its identifier. Communication between devices follows a set of standardized protocols, such as TCP/IP (Transmission Control Protocol/Internet Protocol), to ensure data is transmitted accurately and efficiently.
- **Routing and Network Topology:** Data on the internet travels through a complex network of routers and switches. These devices direct the flow of data based on the most efficient path, considering factors such as latency and network congestion.
- **Domain Name System (DNS):** The DNS translates human-readable domain names (e.g., www.example.com) into IP addresses, allowing users to easily access websites and online services.

3.2.2. Social Media:

- **Algorithms and Machine Learning:** Social media platforms use complex algorithms and machine learning models to personalize content feeds and recommendations for users. These algorithms analyze user interactions and preferences to provide relevant content and improve user engagement.
- **Network Effects and Graph Theory:** Social media platforms thrive on network effects, where the value of the platform increases as more people join and engage. Graph theory is used to model and analyze the relationships and connections within the platform's user network.
- **Data Management and Storage:** Social media platforms generate massive amounts of data, including user-generated content, metadata, and interaction data. Efficient data management and storage systems are essential to handle this volume of information and enable fast retrieval.

3.2.3. Digital Communication:

- **Encoding and Modulation:** Digital communication involves encoding information (e.g., text, images, video) into a digital format (binary code) that can be transmitted over various media such as cables, fiber optics, or wireless signals. Modulation techniques are used to encode the digital signals for transmission.
- **Error Detection and Correction:** During transmission, data can be affected by noise and interference. Error detection and correction techniques, such as parity checks and forward error correction, help ensure the accuracy and integrity of transmitted data.
- **Encryption and Security:** To protect data privacy and confidentiality, digital communication often employs

encryption techniques. Cryptographic algorithms encode data in such a way that only authorized parties can decrypt and access the information.

- **Wireless Communication**: Wireless communication, such as Wi-Fi and mobile networks, uses radio waves to transmit data. Technologies such as 5G enable high-speed wireless communication, providing faster and more reliable connectivity.

The science behind the internet, social media, and digital communication is a blend of information theory, network engineering, and computer science. These technologies work together to create the interconnected, digital world we rely on today. Understanding the science and principles behind these systems can help us appreciate the complexity and innovation that drive our modern communication infrastructure.

Chapter 4: The Mathematics of Chance

The mathematics of chance, also known as probability theory, is a branch of mathematics that deals with the analysis of random events and the likelihood of different outcomes. This field provides a framework for understanding uncertainty and making predictions about events based on known probabilities.

In this chapter, we will explore the key concepts and applications of probability theory. We begin with the basics, such as random variables, probability distributions, and expected values, which form the foundation of the mathematics of chance. By understanding these concepts, we can quantify and model uncertainty in various scenarios.

Next, we will delve into the different types of probability distributions, such as uniform, normal, and binomial distributions, and how they are used to describe the behavior of random variables. These distributions allow us to make inferences and draw conclusions about data and events.

Throughout the chapter, we will examine the applications of probability theory in real-world situations, including risk assessment, decision-making, and statistical analysis. For example, probability theory is used in fields such as finance, insurance, and gambling to calculate odds and manage risk.

We will also explore the relationship between probability and statistics, including how statistical methods use probability theory to analyze data and draw conclusions about populations from sample data.

Finally, we will discuss the philosophical and practical implications of probability theory, including its role in understanding causation, making predictions, and assessing uncertainty. By grasping the mathematics of chance, we gain a deeper understanding of the role of randomness in our lives and how to make informed decisions in the face of uncertainty.

4.1. Understanding probability and its role in everyday events

Probability is a measure of the likelihood of an event occurring, expressed as a number between 0 and 1, where 0 indicates impossibility and 1 indicates certainty. It plays a critical role in our daily lives, from simple activities like flipping a coin to more complex scenarios such as predicting weather patterns or assessing risks in finance and insurance.

4.1.1. The Basics of Probability:

- **Events and Outcomes:** In probability theory, an event is an occurrence or result that can happen, while an outcome is a specific possible result of the event. For example, when rolling a six-sided die, the event is rolling a number, and the outcomes are 1 through 6.
- **Probability Values:** The probability of an event is expressed as a value between 0 and 1. A probability of 0 means the event is impossible, while a probability of 1 means the event is certain to occur. A probability of 0.5 (or 50%) represents equal likelihood of occurrence or non-occurrence.
- **Complementary Events:** The probability of an event occurring and the probability of it not occurring add up to 1. For instance, the probability of rolling an even number on a die is 0.5, and the probability of rolling an odd number is also 0.5.

4.1.2. Probability in Everyday Events:

- **Decision-Making:** Probability helps us make informed decisions by quantifying the likelihood of different outcomes. For instance, when choosing between two travel routes, we might consider the probability of encountering traffic on each one.

- **Gambling and Games of Chance:** In games like poker, blackjack, and roulette, players rely on probability to calculate the odds of winning or losing. Understanding the likelihood of different outcomes can help players make strategic decisions.
- **Weather Forecasting:** Meteorologists use probability to predict weather conditions. For example, when they say there is a 70% chance of rain, they are expressing the likelihood of rain occurring based on data and models.
- **Finance and Investing:** Probability is used in finance to assess the risk and potential return of investments. For example, investors might use probability to estimate the chance of a stock's price increasing or decreasing.
- **Healthcare and Medicine:** Probability plays a role in medical decision-making, such as estimating the likelihood of a patient benefiting from a specific treatment or the chance of a disease recurring.
- **Quality Control and Manufacturing**: In manufacturing, probability is used to assess the quality of products and identify defects. Statistical methods such as control charts help monitor production processes.

4.1.3. Risk Assessment and Management:

- **Insurance:** Insurance companies use probability to calculate premiums based on the likelihood of certain events occurring, such as accidents or natural disasters.
- **Safety and Security:** Probability is used to assess risks in various domains, such as transportation safety and cybersecurity. Understanding the likelihood of different threats allows for the implementation of protective measures.

Probability is a powerful tool that allows us to navigate uncertainty and make informed decisions in our daily lives. By understanding the role of probability in everyday events, we can better assess risks, predict outcomes, and plan for the future.

4.2. The use of statistics in interpreting data and making predictions

Statistics is a branch of mathematics that involves collecting, analyzing, interpreting, presenting, and drawing conclusions from data. It is a powerful tool for making sense of data and making predictions about future events or trends. Statistics plays a crucial role in many aspects of our lives, from informing business decisions to guiding public policy and scientific research.

4.2.1. Collecting and Analyzing Data:

- **Data Collection:** Statistics begins with gathering data, which can be done through surveys, experiments, observations, or other methods. The quality of the data collection process affects the reliability and validity of the subsequent analysis.
- **Descriptive Statistics:** Once data is collected, descriptive statistics summarize and describe the data using measures such as mean, median, mode, variance, and standard deviation. These metrics provide insights into the central tendency, spread, and distribution of the data.

4.2.2. Interpreting Data:

- **Statistical Tests:** Statistical tests are used to determine whether observed data supports a particular hypothesis or theory. Common tests include t-tests, chi-square tests, and analysis of variance (ANOVA), which assess differences between groups or relationships between variables.
- **Confidence Intervals:** Confidence intervals provide a range of values within which a population parameter (such as the mean) is likely to fall, with a certain level of confidence (e.g., 95%).

They offer a way to account for the uncertainty inherent in estimating population parameters from sample data.
- **Regression Analysis:** Regression analysis examines the relationship between one or more independent variables and a dependent variable. This technique can help identify trends and predict future outcomes based on existing data.

4.2.3. Making Predictions:

- **Predictive Models:** Statistical methods, such as linear regression and time series analysis, are used to build models that predict future outcomes based on historical data. These models can be used in various fields, including finance, economics, and healthcare.
- **Machine Learning:** Advanced statistical techniques, such as machine learning algorithms, are increasingly used to analyze large datasets and make predictions. These algorithms can identify complex patterns and relationships in data that traditional statistical methods might miss.
- **Forecasting:** Forecasting uses statistical models to predict future values of a variable based on past data. This is commonly used in business and economics to predict sales, demand, or economic growth.

4.2.4. Applications in Various Fields:

- **Business and Marketing:** Businesses use statistics to analyze consumer behavior, assess market trends, and make strategic decisions about product development and marketing campaigns.

- **Healthcare and Medicine:** In healthcare, statistics is used to evaluate treatment effectiveness, assess risk factors, and inform medical guidelines and public health policies.
- **Sports and Entertainment:** Statistics play a significant role in sports analytics, providing insights into player performance and team strategies. In the entertainment industry, statistics help assess audience preferences and optimize content delivery.
- **Public Policy and Social Sciences:** Governments and researchers use statistics to inform policy decisions, evaluate the effectiveness of programs, and understand social and economic issues.

Statistics is a versatile tool that helps us interpret data, identify patterns, and make informed predictions. By applying statistical methods, we can gain valuable insights into complex datasets and make decisions based on evidence and probability.

Chapter 5: The Physics of Motion

The physics of motion, also known as mechanics, is the branch of physics that studies the behavior of objects in motion. It encompasses both classical mechanics, which deals with the motion of macroscopic objects, and relativistic mechanics, which considers the effects of high speeds close to the speed of light.

In this chapter, we will explore the fundamental concepts of motion, including velocity, acceleration, force, and momentum. We begin with the basic principles of classical mechanics, such as Newton's laws of motion, which describe the relationships between an object's motion and the forces acting on it.

Next, we will examine different types of motion, including linear motion, circular motion, and projectile motion. These concepts help us understand how objects move through space and the forces that influence their trajectories.

Throughout the chapter, we will also discuss the principles of energy, including kinetic and potential energy, and how energy conservation plays a key role in understanding motion. By exploring concepts such as work and power, we can gain insights into how energy is transferred and transformed during motion.

In addition, we will touch on the role of friction and air resistance in motion, as these forces can significantly affect an object's trajectory and speed. Understanding these effects is crucial for accurately predicting and controlling motion in practical applications.

Finally, we will consider the real-world applications of the physics of motion, such as in engineering, transportation, sports, and space exploration. By understanding the principles of motion, we can design more efficient systems, improve safety, and achieve remarkable feats of engineering and technology.

By the end of this chapter, you will have a solid foundation in the physics of motion and an appreciation for how these principles underpin much of the world around us.

5.1. Exploring the principles of motion in everyday activities like sports and transportation

The principles of motion play a crucial role in everyday activities, from sports to transportation. By understanding the physics behind motion, we can improve performance, efficiency, and safety in a variety of settings.

5.1.1. Motion in Sports:

- **Velocity and Acceleration:** Athletes rely on velocity (speed with direction) and acceleration (change in velocity) to perform optimally. For example, in sprinting, an athlete aims to accelerate quickly to reach maximum velocity as fast as possible.
- **Momentum:** In contact sports such as football and hockey, momentum (mass times velocity) is important in understanding collisions and impacts. A larger mass or higher velocity results in greater momentum, affecting the force of a collision.
- **Projectile Motion:** Sports like basketball, soccer, and tennis involve projectile motion, where an object moves along a curved path under the influence of gravity. Players must consider factors such as initial velocity, launch angle, and air resistance when making a pass or taking a shot.
- **Friction:** Friction between athletes' footwear and the ground affects performance and control. For example, in track and field, sprinters benefit from shoes with spikes that increase traction and reduce slipping.
- **Angular Motion:** In sports such as gymnastics, diving, and figure skating, athletes perform complex maneuvers involving rotation. Angular momentum (rotational inertia times angular velocity) plays a key role in controlling spins and flips.

5.1.2. Motion in Transportation

- **Linear Motion:** Cars, trains, and airplanes primarily move in linear paths. Their acceleration and deceleration are controlled to ensure smooth and safe journeys for passengers.
- **Newton's Laws of Motion:** Newton's laws describe how objects move in response to forces. For instance, a car must apply force (via the engine) to overcome inertia and start moving, and friction (from brakes) to stop.
- **Friction and Air Resistance:** Friction between a vehicle's tires and the road affects traction and maneuverability. Air resistance, also known as drag, impacts a vehicle's speed and fuel efficiency, especially at high velocities.
- **Energy Efficiency:** Energy considerations, such as minimizing fuel consumption, are important in transportation. Electric and hybrid vehicles are designed to be energy-efficient by converting potential energy from batteries into kinetic energy.
- **Circular Motion:** In transportation systems like trains on curved tracks and cars on roundabouts, circular motion principles are applied to navigate curves safely. Centripetal force, directed toward the center of the circle, is crucial to keep vehicles on track.
- **Momentum and Safety:** In transportation, momentum plays a role in safety during collisions. Seatbelts and airbags are designed to counteract the forces experienced by passengers in a collision, reducing injuries.

Understanding the principles of motion helps athletes and engineers optimize performance, safety, and efficiency in sports and transportation. By applying these concepts, we can develop better equipment, improve training methods, and design safer and more efficient transportation systems.

5.2. The role of physics in explaining why things move the way they do

Physics provides a framework for understanding the principles governing the movement of objects, known as mechanics. These principles help explain why things move the way they do, considering factors such as forces, energy, and momentum. Here are some key concepts from physics that explain motion:

5.2.1. Newton's Laws of Motion:

- **First Law (Law of Inertia):** An object at rest will stay at rest, and an object in motion will stay in motion at a constant velocity, unless acted upon by an external force. This explains why objects continue in their state of motion unless a force (e.g., friction, gravity) intervenes.
- **Second Law (Law of Acceleration):** The acceleration of an object is directly proportional to the net force acting on it and inversely proportional to its mass (F=ma). This law helps us understand how forces such as gravity, friction, and applied forces affect an object's motion.
- **Third Law (Law of Action and Reaction):** For every action, there is an equal and opposite reaction. This law explains interactions such as when a car pushes against the road to move forward, and the road pushes back with an equal force.

5.2.2. Energy and Work:

- **Kinetic Energy:** The energy of an object due to its motion is known as kinetic energy. It is given by the formula is KE = 1/2 mv² velocity. Kinetic energy helps explain the effect of an object's speed and mass on its ability to perform work.

- **Potential Energy:** Potential energy is the energy stored in an object due to its position or state. For example, gravitational potential energy (PE=mgh) depends on an object's height above the ground and its mass.
- **Conservation of Energy:** The total energy in an isolated system remains constant, though it can change forms (e.g., from potential to kinetic energy). This principle helps explain how energy is transferred and transformed during motion.

5.2.3. Momentum and Impulse:

- **Momentum:** Momentum (p=mv) is the product of an object's mass and velocity. It represents the quantity of motion an object possesses and is conserved in an isolated system.
- **Impulse:** Impulse is the change in momentum caused by a force acting on an object over a period of time (J=F·t). This concept helps explain how forces cause changes in an object's motion.

5.2.4. Friction and Air Resistance:

- **Friction:** Friction is a force that opposes an object's motion due to contact with another surface. It plays a key role in stopping or slowing down objects and is influenced by the nature of the surfaces in contact.
- **Air Resistance:** Also known as drag, air resistance is a force that opposes an object's motion through the air. It depends on factors such as the object's shape, speed, and the density of the air.

5.2.5. Circular and Rotational Motion:

- **Centripetal Force:** Objects moving in circular paths experience a centripetal force directed toward the center of the circle. This force keeps the object moving along the curved path.
- **Angular Momentum:** Angular momentum ($L=I\omega$) is the product of an object's moment of inertia and its angular velocity. It is conserved in an isolated system and helps explain the stability and motion of rotating objects.

Physics provides a comprehensive explanation of why things move the way they do by combining concepts such as forces, energy, momentum, and rotational motion. Understanding these principles helps us predict, analyze, and control the motion of objects in everyday life and in advanced technological applications.

Chapter 6: The IT in Everyday Devices

In today's world, information technology (IT) is deeply integrated into our everyday lives, powering the devices we use daily. From smartphones and smart home devices to wearable technology and personal computers, IT is at the core of the functionality and connectivity of these devices. This chapter explores the IT in everyday devices, focusing on the key components, technologies, and applications that enable these devices to function and enhance our lives.

Key Components of Everyday Devices:

- **Microprocessors and CPUs:** The central processing unit (CPU) is the "brain" of a device, executing instructions and performing calculations. Microprocessors integrate multiple functions, including the CPU, memory, and peripherals, into a single chip.
- **Memory:** Devices use different types of memory for data storage, including volatile memory (e.g., RAM) for temporary data and non-volatile memory (e.g., flash memory) for long-term data storage.
- **Sensors and Actuators:** Sensors collect data from the environment (e.g., temperature, light, motion) and convert it into digital signals for processing. Actuators, on the other hand, perform actions based on processed data, such as adjusting settings or moving components.
- **Displays and Interfaces:** Displays (e.g., LCD, OLED) allow users to view information, while interfaces such as touchscreens and buttons enable user interaction with the device.

<u>*Connectivity and Networking:*</u>

- **Wi-Fi and Bluetooth:** Most everyday devices use Wi-Fi and Bluetooth for wireless connectivity, allowing them to communicate with other devices and access the internet.
- **IoT (Internet of Things):** Many everyday devices are part of the IoT ecosystem, enabling them to connect and share data with other devices over the internet. This interconnectivity allows for smart homes and connected appliances.

<u>*Software and Operating Systems:*</u>

- **Operating Systems:** Everyday devices run operating systems (e.g., iOS, Android, Windows) that manage hardware resources, provide a user interface, and enable the execution of software applications.
- **Apps and Software Applications:** Devices use software applications (apps) to provide functionality and services. For example, apps on a smartphone enable communication, navigation, and entertainment.

<u>*Security and Privacy:*</u>

- **Data Security:** Everyday devices store and transmit personal data, making security a priority. Encryption and secure protocols protect data from unauthorized access.
- **Privacy Controls:** Devices offer privacy settings to control data sharing and permissions, allowing users to manage their digital footprint.

<u>*Power Management and Efficiency:*</u>

- **Battery Technology:** Many everyday devices rely on rechargeable batteries for power. Advances in battery technology improve energy density, charging speed, and lifespan.
- **Energy Efficiency:** Devices incorporate energy-saving features such as sleep modes and adaptive brightness to extend battery life and reduce power consumption.

Applications in Everyday Life:

- **Smartphones:** Smartphones are versatile devices used for communication, internet access, navigation, photography, and more.
- **Wearable Technology:** Wearable devices, such as smartwatches and fitness trackers, monitor health and fitness metrics and provide notifications.
- **Smart Home Devices:** Smart home devices, such as thermostats, lighting, and security systems, allow users to control their home environment remotely.
- **Voice Assistants:** Voice-activated assistants, like Amazon Alexa and Google Assistant, provide hands-free control of devices and access to information.

Information technology in everyday devices enhances our lives by providing connectivity, convenience, and access to a wealth of information and services. As IT continues to advance, we can expect further integration of intelligent, connected devices into our daily routines.

6.1. Understanding how IT powers common devices like smartphones and home appliances

Information technology (IT) plays a crucial role in powering common devices such as smartphones and home appliances, enabling them to function effectively and offer a variety of features that enhance our everyday lives. Let's explore how IT powers these devices:

6.1.1. Smartphones:

- **Processors and Chipsets:** Smartphones are equipped with processors and chipsets that handle data processing, communication, and graphics rendering. Modern smartphones often use System-on-Chip (SoC) technology, which integrates the CPU, GPU (graphics processing unit), and other components into a single chip.
- **Operating Systems:** Smartphones run on operating systems such as iOS and Android, which manage hardware resources and provide a user interface. The operating system also facilitates app management and security.
- **Memory and Storage:** Smartphones use RAM (random-access memory) for temporary data storage during operation and flash memory for long-term storage of data such as photos, videos, and apps.
- **Sensors and Cameras:** Smartphones come with a variety of sensors (e.g., accelerometer, gyroscope, proximity, ambient light) that enable features such as auto-rotation and ambient light adjustments. Cameras on smartphones incorporate advanced sensors for high-quality photography and video recording.
- **Connectivity:** Smartphones connect to the internet and other devices using Wi-Fi and mobile data networks (e.g., 4G, 5G).

Bluetooth allows smartphones to connect wirelessly to peripherals such as headphones and speakers.

- **Apps:** Apps provide a range of functions on smartphones, from communication and entertainment to productivity and navigation. App stores make it easy for users to download and update apps.
- **Security and Privacy:** Smartphones use encryption and secure protocols to protect user data. Operating systems offer privacy controls for managing app permissions and data sharing.

6.1.2. Home Appliances:

- **Microcontrollers and Processors:** Home appliances often use microcontrollers and processors to manage functions such as temperature control, timing, and sensor data processing. These components enable automation and efficiency in appliance operation.
- **Connectivity and IoT:** Many modern home appliances, such as smart thermostats, refrigerators, and washing machines, are part of the Internet of Things (IoT) ecosystem. They connect to the internet via Wi-Fi or other protocols, allowing for remote control and monitoring through smartphone apps.
- **Sensors and Actuators:** Home appliances use sensors to detect changes in the environment (e.g., temperature, humidity, load weight) and actuators to perform actions based on sensor data (e.g., adjusting temperature, turning on water supply).
- **Interfaces and Displays:** Appliances may have touchscreens, buttons, or dials for user input, as well as displays for providing information such as temperature or cycle status.
- **Energy Efficiency:** IT helps home appliances operate more efficiently through features such as smart scheduling, adaptive heating or cooling, and energy usage monitoring.

- **Voice Control:** Many smart home appliances can be controlled using voice assistants such as Amazon Alexa and Google Assistant. This allows for convenient, hands-free operation.

6.1.3. Applications in Everyday Life:

- **Remote Control and Monitoring:** IT enables users to control and monitor home appliances remotely through smartphone apps. For example, a user can adjust the thermostat, start a load of laundry, or check the contents of a smart fridge from their phone.
- **Automation and Personalization:** IT allows for automation and personalization of home appliance functions. For example, smart lighting can adjust brightness and color temperature based on the time of day and user preferences.
- **Safety and Security:** IT-powered appliances can include safety features such as automatic shut-off for ovens and alerts for water leaks in washing machines.

Information technology is integral to the functionality and versatility of common devices like smartphones and home appliances. By combining hardware, software, and connectivity, IT enhances the capabilities of these devices and makes our lives more convenient and efficient.

6.2. The science behind computer hardware and software

Computer hardware and software are the fundamental building blocks of information technology. Together, they enable computing devices to perform a wide range of tasks, from data processing to running complex applications. Let's explore the science behind computer hardware and software and how they work together to power modern devices.

6.2.1. Computer Hardware:

Computer hardware refers to the physical components of a computing device. Key hardware components include:

- **Central Processing Unit (CPU):** The CPU is the "brain" of the computer. It executes instructions from software and performs calculations, logic operations, and data processing. CPUs contain multiple cores, which allow for parallel processing and multitasking.
- **Memory:** Memory (RAM) is used for temporary data storage during program execution. The amount of RAM affects a computer's ability to handle multiple tasks simultaneously.
- **Storage:** Storage devices, such as hard drives (HDD) and solid-state drives (SSD), provide long-term storage of data, including operating systems, applications, and files. SSDs are faster and more reliable than traditional HDDs.
- **Graphics Processing Unit (GPU):** GPUs handle graphics rendering and are essential for tasks such as video editing, gaming, and artificial intelligence (AI) applications.
- **Motherboard:** The motherboard is the main circuit board that connects and coordinates communication between all hardware components.

- **Peripherals:** Peripherals include input devices (e.g., keyboard, mouse), output devices (e.g., monitor, printer), and other devices that connect to the computer.

6.2.2. Computer Software:

Computer software refers to the programs and applications that run on hardware. Software can be categorized into:

- **Operating Systems:** The operating system (e.g., Windows, macOS, Linux) manages hardware resources, provides a user interface, and enables the execution of software applications. It acts as an intermediary between hardware and software.
- **Applications:** Applications, also known as apps or software programs, perform specific tasks for users. They can be categorized into productivity (e.g., word processors, spreadsheets), communication (e.g., email, messaging), entertainment (e.g., games, media players), and more.
- **Programming Languages:** Software is written using programming languages such as Python, Java, C++, and JavaScript. These languages provide syntax and rules for writing code that the computer can execute.
- **Compilers and Interpreters:** Compilers and interpreters translate high-level programming languages into machine code that the CPU can understand and execute.
- **Software Development Tools:** Tools such as integrated development environments (IDEs), debuggers, and version control systems help developers create, test, and maintain software.

6.2.3. The Interaction Between Hardware and Software:

- **Instruction Execution**: Software provides instructions for the hardware to execute. The CPU fetches instructions from memory and processes them, coordinating with other hardware components as needed.
- **Device Drivers:** Device drivers are software that allow the operating system to communicate with hardware peripherals (e.g., printers, scanners, graphics cards).
- **Firmware:** Firmware is low-level software that controls the hardware directly. It is stored in read-only memory (ROM) and initializes hardware components when the device is powered on.

6.2.4. Advancements in Hardware and Software:

- **Parallel Processing and Multithreading:** Modern CPUs and GPUs support parallel processing and multithreading, allowing for efficient execution of multiple tasks simultaneously.
- **Artificial Intelligence and Machine Learning:** Hardware and software advancements have enabled the rise of AI and machine learning applications, which require powerful processing capabilities.
- **Cloud Computing:** Cloud computing leverages both hardware and software to provide on-demand access to computing resources, such as storage and processing power, over the internet.

The interplay between computer hardware and software drives the capabilities of modern computing devices. By understanding the science behind these components, we can appreciate how they work together to power a vast array of digital technologies and applications in our daily lives.

Chapter 7: The Mathematics of Nature

Nature is filled with patterns and structures that can be described and understood through mathematics. From the symmetrical arrangements of leaves and flowers to the complex behaviors of weather systems and ecosystems, mathematical principles govern many aspects of the natural world.

In this chapter, we will explore the mathematics of nature, examining how mathematical concepts help us understand and describe the patterns, structures, and dynamics of natural phenomena. We will begin by discussing the geometric shapes and arrangements that occur in nature, such as the Fibonacci sequence, golden ratio, and fractals.

Next, we will delve into the mathematical modeling of natural systems, such as population dynamics, predator-prey relationships, and the spread of diseases. These models help us predict and analyze the behavior of complex biological and ecological systems.

We will also explore the role of mathematics in understanding physical phenomena such as fluid dynamics, which describes the flow of air and water, and thermodynamics, which explains the transfer of energy in natural processes.

Throughout the chapter, we will touch on the use of mathematics in fields such as astronomy, where mathematical equations describe the movements of celestial bodies, and geology, where mathematical models help explain the formation and movement of geological structures.

Finally, we will consider the impact of mathematical understanding on our ability to interact with and protect the natural world. By applying mathematical principles, we can make informed decisions about resource management, conservation, and sustainability.

By the end of this chapter, readers will have a greater appreciation for the mathematics of nature and its role in revealing the intricate beauty and complexity of the world around us. Understanding these mathematical concepts allows us to see the interconnectedness of natural systems and helps us navigate and preserve the delicate balance of our environment.

7.1. Exploring patterns and sequences in the natural world

Nature is replete with patterns and sequences that can be described and analyzed using mathematical concepts. These patterns range from the shapes and arrangements of leaves and flowers to the spirals and symmetries found in shells and galaxies. By studying these patterns and sequences, we can gain insights into the underlying processes and principles that govern the natural world.

7.1.1. Fibonacci Sequence:

One of the most famous sequences in nature is the Fibonacci sequence, a series of numbers where each number is the sum of the two preceding ones (0, 1, 1, 2, 3, 5, 8, 13, …). This sequence appears in many natural structures, such as the arrangement of petals on a flower, the branching of trees, and the spirals of shells.

The Fibonacci sequence often gives rise to aesthetically pleasing forms, which is why it is associated with concepts such as the golden ratio (approximately 1.618). The golden ratio is seen in the proportions of certain plants, animals, and even human features, as well as in art and architecture.

7.1.2. Fractals:

Fractals are complex patterns that repeat at different scales, often resulting in shapes that resemble each other at various levels of magnification. In nature, fractals can be seen in the branching of trees, the formation of clouds, the structure of coastlines, and the arrangement of mountain ranges.

These patterns arise from iterative processes where simple rules are applied repeatedly. Fractals help us understand the complexity of natural systems and their self-similar structures.

7.1.3. Symmetry:

Symmetry is another important pattern found in nature. It can be seen in the bilateral symmetry of animals, such as butterflies and humans, and the radial symmetry of flowers and starfish. Symmetry often serves a functional purpose, such as aiding in movement, reproduction, or defense.

Mathematics helps us describe and classify different types of symmetry, such as translational, rotational, and reflection symmetry. These classifications provide a framework for understanding the geometric properties of natural objects.

7.1.4. Spirals and Waves:

Spirals and waves are recurring patterns in nature that can be described mathematically. Spirals, such as those found in seashells and galaxies, follow specific mathematical relationships, such as the logarithmic spiral.

Waves, such as sound waves, light waves, and water waves, follow mathematical principles that describe their frequency, amplitude, and wavelength. These patterns help explain phenomena such as the transmission of energy and the propagation of signals.

7.1.5. Tiling and Packing:

The arrangement of objects in a space can be analyzed using mathematical concepts such as tiling and packing. For example, the honeycomb structure of a beehive represents an efficient packing arrangement that maximizes storage while minimizing material usage.

By exploring these and other patterns and sequences in the natural world, we can better understand the underlying principles that shape the environment around us. Mathematics provides a powerful language for describing and analyzing these phenomena, revealing the beauty and complexity of nature.

7.2. The role of mathematics in understanding phenomena like fractals and the Fibonacci sequence

Mathematics plays a pivotal role in understanding and describing complex phenomena in nature such as fractals and the Fibonacci sequence. These mathematical concepts offer insights into the patterns and structures that pervade the natural world, providing a framework for analyzing and appreciating the intricate designs found in plants, animals, and other natural systems.

7.2.1. Fractals:

Fractals are patterns that exhibit self-similarity at different scales, meaning that smaller sections of the pattern resemble the whole. These patterns are often generated by iterative processes that apply simple rules repeatedly. Mathematicians have developed various methods for creating and analyzing fractals, which helps us understand their properties and the natural phenomena they model.

- **Generation of Fractals:** Fractals can be generated using recursive algorithms that apply transformations to a shape or pattern repeatedly. Examples include the Mandelbrot set and the Sierpinski triangle.
- **Properties of Fractals:** Fractals often have a non-integer or fractional dimension, reflecting their complexity and the fact that they occupy more space than a line but less than a plane.
- **Applications in Nature:** Fractals appear in many natural systems, such as the branching patterns of trees, rivers, and blood vessels, as well as the jagged shapes of coastlines and mountains. Mathematical analysis of fractals helps explain the efficiency and adaptability of these natural structures.

7.2.2. Fibonacci Sequence:

The Fibonacci sequence is a series of numbers where each number is the sum of the two preceding ones (0, 1, 1, 2, 3, 5, 8, 13, ...). This sequence and its related concepts, such as the golden ratio, appear in a variety of natural phenomena.

- **Mathematical Properties:** The ratio of successive Fibonacci numbers converges to the golden ratio (approximately 1.618), a constant that appears in various aspects of geometry and art.
- **Applications in Nature:** The Fibonacci sequence and the golden ratio are found in many natural patterns, such as the arrangement of leaves, petals, and seeds in plants. This arrangement allows plants to efficiently capture sunlight and distribute resources.
- **Modeling Growth:** The Fibonacci sequence can model growth patterns in nature, such as population growth in certain species or the proliferation of cells.

By using mathematical tools to analyze phenomena like fractals and the Fibonacci sequence, we can gain a deeper understanding of the underlying principles and processes that drive the organization and growth of natural systems. Mathematics provides a common language for describing these patterns, allowing us to appreciate their beauty and complexity while also uncovering their functional significance in nature.

Chapter 8: The Physics of Sound

Sound is an essential aspect of our world, allowing us to communicate, perceive our environment, and experience music and other forms of art. The physics of sound explores how sound is produced, transmitted, and perceived, as well as its interaction with different materials and environments.

In this chapter, we will explore the fundamental concepts of sound, including its nature as a mechanical wave that propagates through various media such as air, water, and solids. We will discuss the properties of sound waves, such as frequency, wavelength, amplitude, and speed, and how these properties influence the characteristics of sound, including pitch, loudness, and timbre.

Next, we will delve into the mechanisms of sound production, transmission, and perception. This includes the generation of sound waves by vibrating sources, the transmission of sound through different media, and the reception and interpretation of sound by the human ear and brain.

Throughout the chapter, we will examine how sound interacts with its environment, including reflection, refraction, diffraction, and absorption. These interactions give rise to phenomena such as echoes, reverberation, and the Doppler effect.

We will also explore the various applications of sound in technology, including acoustics, audio engineering, and noise control. These applications have practical implications in fields such as music production, architecture, medicine, and communication.

Finally, we will consider the impact of sound on our daily lives, including the role of sound in shaping our experiences and well-being. By understanding the physics of sound, we gain insight into how to manipulate and control sound effectively, allowing us to enhance our auditory experiences and mitigate unwanted noise.

Let's begin our exploration of the physics of sound and its significance in the world around us.

8.1. Understanding the nature of sound and its role in communication and music

Sound is a form of mechanical wave that travels through different media, such as air, water, and solids. It is produced by vibrating objects, which create pressure waves that propagate through the medium. These pressure waves consist of compressions and rarefactions, which move away from the source of the vibration.

8.1.1. The Nature of Sound:

Sound waves can be characterized by several key properties:

- **Frequency:** The frequency of a sound wave is the number of oscillations (vibrations) that occur per second, measured in hertz (Hz). Frequency determines the pitch of the sound, with higher frequencies corresponding to higher pitches.
- **Wavelength:** The wavelength is the distance between consecutive compressions or rarefactions in the wave. It is inversely related to frequency; higher frequency waves have shorter wavelengths.
- **Amplitude:** The amplitude of a sound wave is the magnitude of the pressure change caused by the wave. Amplitude determines the loudness or intensity of the sound.
- **Speed:** The speed of sound varies depending on the medium through which it travels. In air at room temperature, sound travels at approximately 343 meters per second (1,125 feet per second).

8.1.2. Sound in Communication:

Sound plays a vital role in human communication, particularly through spoken language. The human vocal cords produce sound waves that are modulated by the mouth, tongue, and lips to create speech. These sound waves carry information that is interpreted by the listener's auditory system.

Advancements in technology have expanded the ways in which sound is used for communication. Telephones, microphones, and speakers convert sound waves into electrical signals and vice versa, allowing for long-distance and amplified communication.

8.1.3. Sound in Music:

Music is a form of artistic expression that relies heavily on sound. Musical sounds are produced by instruments or vocal cords, and they are characterized by specific frequencies, amplitudes, and durations. Music often involves the manipulation of these elements to create harmony, melody, and rhythm.

- **Timbre:** Timbre, or the quality of sound, allows us to distinguish between different musical instruments or voices. It is determined by the complex mix of frequencies and harmonics present in a sound.
- **Harmony and Melody:** Harmony involves the combination of different musical notes to create chords, while melody is a sequence of notes that forms a recognizable tune.
- **Rhythm:** Rhythm refers to the timing and patterns of sounds in music, including the duration and spacing of notes and beats.

Musicians and audio engineers use the principles of acoustics to optimize the production and performance of music. This includes adjusting factors such as sound levels, reverberation, and frequency balance to achieve the desired auditory experience.

In summary, sound is an essential element of both communication and music. By understanding its nature and properties, we can appreciate how sound enhances our ability to share information and express ourselves artistically. Moreover, advancements in sound technology have further enriched our experiences in these areas.

8.2. The science behind acoustics and noise cancellation

Acoustics is the branch of physics that studies the properties of sound and its interactions with different environments and materials. The field encompasses the production, transmission, and reception of sound waves, as well as their control and manipulation. One of the key applications of acoustics is noise cancellation, which involves reducing unwanted sounds to enhance listening experiences and protect hearing.

8.2.1. Acoustics:

Acoustics involves understanding how sound behaves in various settings, including how it travels through different media and interacts with surfaces and structures. Key concepts in acoustics include:

- **Reflection:** Sound waves can bounce off surfaces, creating echoes. The angle of incidence (the angle at which the sound wave strikes the surface) is equal to the angle of reflection.
- **Absorption:** Materials can absorb sound energy, converting it into other forms of energy, such as heat. Soft, porous materials are effective at absorbing sound.
- **Refraction:** Sound waves can bend as they pass through different media or areas of varying temperature and density. This bending is called refraction.
- **Diffraction:** Sound waves can bend around obstacles and spread out through openings. This behavior allows sound to travel beyond barriers and fill large spaces.
- **Reverberation:** Reverberation occurs when sound waves reflect multiple times off surfaces in an enclosed space. This

can create an echo-like effect, which can enhance or detract from the listening experience depending on the context.

Acoustic engineers use these principles to design spaces such as concert halls, recording studios, and theaters, optimizing the quality of sound for specific purposes.

8.2.2. Noise Cancellation:

Noise cancellation, or active noise control, is a technology used to reduce or eliminate unwanted sounds. It works by creating sound waves that are the exact opposite (or "out of phase") of the unwanted noise, effectively canceling it out. There are two main approaches to noise cancellation:

- **Active Noise Cancellation (ANC):** ANC technology uses microphones to pick up ambient noise and then generates sound waves with the opposite phase to the unwanted noise. These canceling sound waves are played back through speakers or headphones, effectively reducing the perceived noise.
- **Passive Noise Isolation:** Unlike ANC, passive noise isolation relies on physical barriers to block or reduce sound transmission. For example, over-ear headphones and earplugs create a seal around the ears to reduce external noise.

Noise cancellation has a variety of applications, from enhancing the audio experience in consumer electronics (e.g., headphones) to improving the performance of pilots and drivers who rely on clear communication and concentration.

By understanding the science behind acoustics and noise cancellation, engineers and designers can create environments and devices that provide optimal sound quality and reduce the impact of noise pollution. This has practical benefits for communication, entertainment, and overall well-being.

Chapter 9: The IT Behind Artificial Intelligence

Artificial Intelligence (AI) has rapidly become an integral part of modern technology, revolutionizing various industries and reshaping our world. At its core, AI relies on sophisticated information technology (IT) systems to process vast amounts of data, learn from it, and make intelligent decisions. In this chapter, we will explore the IT infrastructure and technologies that underpin AI, providing the computational power and data management necessary for its development and deployment.

First, we will discuss the hardware infrastructure required for AI, including powerful processors and specialized hardware such as graphics processing units (GPUs) and tensor processing units (TPUs). These components enable AI algorithms to perform complex computations at high speeds, supporting machine learning and deep learning models.

Next, we will delve into the software frameworks and libraries that provide the foundation for AI development. These tools offer pre-built algorithms and efficient data processing capabilities, streamlining the creation and training of AI models.

The chapter will also explore the role of data in AI, emphasizing the importance of data storage, management, and processing. AI systems rely on large datasets for training, making data storage and retrieval essential aspects of the IT behind AI. We will discuss various data management strategies and technologies, including distributed storage systems and data pipelines.

Security and ethical considerations are crucial when it comes to AI and its IT infrastructure. We will address the need for secure and robust systems to protect sensitive data and ensure the ethical use of AI technologies.

Finally, we will discuss the role of cloud computing in AI, which has become a key enabler for scalable and accessible AI solutions. Cloud platforms provide the necessary resources and infrastructure for developing, deploying, and managing AI applications.

By understanding the IT behind artificial intelligence, we gain insight into how AI systems are built, maintained, and optimized. This knowledge is essential for both AI practitioners and those who wish to understand the technological backbone of this transformative field.

9.1. Exploring the principles of artificial intelligence and machine learning

Artificial Intelligence (AI) is a branch of computer science that focuses on creating machines capable of performing tasks that typically require human intelligence. These tasks may include understanding natural language, recognizing patterns, making decisions, solving problems, and learning from experience. Within AI, machine learning (ML) is a subfield that plays a crucial role in enabling systems to learn and improve their performance over time.

9.1.1. Principles of Artificial Intelligence:

AI encompasses a wide range of techniques and approaches that aim to replicate or simulate intelligent behavior. Some key principles include:

- **Perception:** AI systems can interpret sensory inputs such as visual data from images or videos and auditory data from sound recordings. Techniques like computer vision and speech recognition allow AI to understand and interact with the world.
- **Reasoning and Problem-Solving:** AI systems use algorithms and logic to solve complex problems. This can involve deductive reasoning, where conclusions are drawn from given premises, or inductive reasoning, where patterns and trends are inferred from data.
- **Planning:** AI can generate plans and sequences of actions to achieve specific goals. Planning algorithms help systems navigate complex environments and make decisions about how to proceed.
- **Natural Language Processing (NLP):** NLP enables AI to understand and generate human language, allowing for interactions with users through text or speech.

- **Knowledge Representation:** AI systems often use data structures and algorithms to represent knowledge about the world, enabling them to understand relationships between entities and concepts.

9.1.2. Machine Learning Fundamentals:

Machine learning is a subset of AI that focuses on developing algorithms and models that allow computers to learn from data and improve their performance over time without being explicitly programmed. Key concepts in machine learning include:

- **Supervised Learning:** In supervised learning, models are trained on labeled data, meaning each training example includes input data and the corresponding desired output. The model learns to map inputs to outputs, and the training process involves adjusting model parameters to minimize the error between predicted and actual outputs.
- **Unsupervised Learning:** In unsupervised learning, models are trained on unlabeled data, without explicit instructions on what the output should be. The goal is to identify patterns, groupings, or relationships in the data. Techniques such as clustering and dimensionality reduction fall under this category.
- **Reinforcement Learning:** In reinforcement learning, an AI agent learns by interacting with an environment and receiving feedback in the form of rewards or penalties. The agent aims to learn a policy that maximizes cumulative rewards over time.
- **Deep Learning:** Deep learning is a subset of machine learning that uses artificial neural networks with many layers (deep neural networks) to model complex relationships in data. Deep learning has enabled significant advancements in areas such as image recognition, natural language understanding, and game playing.

Machine learning's ability to adapt and improve based on data has made it a powerful tool in various applications, including predictive analytics, recommendation systems, autonomous vehicles, and more.

9.1.3. Types of Machine Learning Models:

Machine learning encompasses a variety of model types, each with its own strengths and applications:

- **Regression Models:** These models are used to predict continuous outcomes based on input variables. Linear regression is a simple example, while more complex methods like polynomial regression and support vector regression offer greater flexibility.
- **Classification Models:** Classification involves categorizing input data into discrete classes. Common classification models include logistic regression, decision trees, random forests, and support vector machines.
- **Clustering Models:** Clustering involves grouping similar data points together without predefined labels. K-means clustering and hierarchical clustering are popular methods for discovering natural groupings in data.
- **Dimensionality Reduction:** High-dimensional data can be challenging to work with, so dimensionality reduction techniques like principal component analysis (PCA) and t-distributed stochastic neighbor embedding (t-SNE) help simplify and visualize complex data.
- **Ensemble Methods:** Ensemble methods combine multiple models to improve prediction accuracy and robustness. Examples include bagging (e.g., random forests) and boosting (e.g., gradient boosting machines).

- **Sequence Models:** These models handle data with temporal dependencies, such as time series or natural language. Recurrent neural networks (RNNs) and long short-term memory (LSTM) networks are commonly used for tasks like speech recognition and language translation.

9.1.4. The Training and Testing Process:

The process of training a machine learning model involves feeding it a dataset and adjusting its parameters based on how well it predicts the output. This training process typically involves an optimization algorithm, such as gradient descent, which minimizes the difference between predicted and actual values.

Once the model is trained, it is tested on a separate set of data that it has not seen before (testing set) to evaluate its performance and generalization capabilities. Metrics such as accuracy, precision, recall, and F1 score are used to assess the model's performance.

9.1.5. Ethical Considerations in AI and ML:

As AI and machine learning become more prevalent, ethical considerations have taken on increasing importance. Developers must ensure that their models are fair, transparent, and unbiased:

- **Bias and Fairness:** AI systems can inadvertently perpetuate existing biases present in training data, leading to unfair treatment or discrimination. Developers must take steps to detect and mitigate bias in models.
- **Transparency:** AI models, particularly deep learning models, can sometimes be opaque in their decision-making processes. Efforts to enhance model interpretability can help increase trust and accountability.

- **Privacy:** Machine learning often relies on large datasets that may contain sensitive information. Ensuring data privacy through secure data handling and anonymization is essential.

By adhering to ethical standards and best practices, AI and machine learning can be developed responsibly, benefiting society while minimizing potential risks.

Overall, AI and machine learning represent powerful technologies that are transforming numerous industries and shaping the future. Understanding their principles and applications is essential for leveraging their potential while navigating the challenges they present.

Understanding the principles of AI and machine learning is crucial for developing intelligent systems that can learn, adapt, and perform complex tasks, leading to advancements in technology and transforming industries worldwide.

9.2. The role of IT in powering modern AI applications

Information Technology (IT) plays a crucial role in powering modern artificial intelligence (AI) applications. It provides the necessary infrastructure, tools, and resources that enable AI to be developed, trained, and deployed effectively and efficiently. From high-performance hardware to cloud-based services and data management systems, IT underpins the functioning of AI in today's world.

9.2.1. High-Performance Computing and Hardware:

- **Processors and Accelerators:** AI models, particularly deep neural networks, require significant computational power. Modern AI relies on high-performance computing hardware, such as graphics processing units (GPUs) and tensor processing units (TPUs), to accelerate training and inference.
- **Distributed Computing:** Training AI models on large datasets can be resource-intensive and time-consuming. Distributed computing techniques allow the workload to be split across multiple processors or machines, reducing the training time and enabling scalability.
- **Edge Computing:** In some applications, AI processing needs to be done locally on devices (e.g., smartphones, IoT devices) for latency, privacy, or connectivity reasons. Edge computing enables AI models to run efficiently on resource-constrained devices.

9.2.2. Software Frameworks and Libraries:

- **AI Frameworks:** Modern AI applications rely on software frameworks such as TensorFlow, PyTorch, and Keras, which provide tools and libraries for building, training, and deploying

AI models. These frameworks offer pre-built algorithms and efficient data processing capabilities, making AI development more accessible and efficient.
- **API and SDKs:** Application programming interfaces (APIs) and software development kits (SDKs) allow developers to integrate AI capabilities into their applications seamlessly. They provide standardized interfaces for tasks such as image recognition, natural language processing, and predictive analytics.

9.2.3. Data Management and Processing:

- **Data Storage and Retrieval:** AI relies on large datasets for training and evaluation. IT infrastructure such as distributed databases and data lakes enables efficient storage, retrieval, and management of data.
- **Data Preprocessing:** Raw data often needs to be cleaned, transformed, and normalized before being used to train AI models. IT tools and pipelines automate these data preprocessing tasks, ensuring that the data is suitable for AI.
- **Data Privacy and Security:** Ensuring the privacy and security of data used in AI applications is crucial. IT systems provide mechanisms for data encryption, access control, and compliance with data protection regulations.

9.2.4. Cloud Computing and AI-as-a-Service:

- **Cloud Infrastructure:** Cloud computing platforms provide scalable and flexible infrastructure for AI development and deployment. Cloud resources can be provisioned on-demand, allowing developers to access high-performance computing and storage without significant upfront investment.

- **AI-as-a-Service:** Many cloud platforms offer AI-as-a-Service, providing pre-trained models and AI tools that can be easily integrated into applications. This enables developers to leverage AI capabilities without building models from scratch.

9.2.5. Deployment and Monitoring:

- **Model Deployment:** Once an AI model is trained, it needs to be deployed in a production environment for real-world use. IT infrastructure supports the deployment of AI models as services that can be accessed via APIs.
- **Monitoring and Maintenance:** IT systems are used to monitor the performance of AI models in production, ensuring they operate efficiently and accurately. Continuous monitoring helps detect issues such as model drift, where model accuracy may degrade over time.

In summary, IT is foundational to modern AI applications, providing the hardware, software, data management, and infrastructure necessary for AI development, deployment, and maintenance. As AI continues to evolve, IT will remain a key driver in advancing AI capabilities and expanding its impact across industries.

Chapter 10: The Chemistry of Life

The chemistry of life, or biochemistry, is the study of the chemical processes and substances that sustain life in living organisms. It explores the molecular mechanisms underlying vital biological functions such as metabolism, replication, and cellular communication. Understanding the chemistry of life allows us to appreciate how complex molecular interactions enable living systems to grow, develop, and respond to their environment.

In this chapter, we will delve into the foundational principles of biochemistry, starting with the basic building blocks of life: carbon-based molecules such as carbohydrates, lipids, proteins, and nucleic acids. These macromolecules play essential roles in the structure and function of cells, tissues, and organs.

Next, we will explore the intricate metabolic pathways that enable organisms to convert nutrients into energy and build the molecules necessary for growth and maintenance. We will discuss the importance of enzymes, which act as biological catalysts to speed up chemical reactions and regulate metabolic processes.

The chapter will also examine the role of chemical signaling in life, highlighting the significance of hormones, neurotransmitters, and other signaling molecules in coordinating bodily functions and facilitating communication between cells.

We will explore the concept of molecular genetics, which focuses on the role of DNA and RNA in storing, transmitting, and expressing genetic information. This knowledge is crucial for understanding how traits are inherited and how genetic variations can lead to diseases.

Finally, we will discuss the applications of biochemistry in medicine, agriculture, and biotechnology, illustrating how our understanding of the chemistry of life has led to advances in drug development, diagnostics, and genetic engineering.

By understanding the chemistry of life, we gain insight into the molecular basis of living organisms and the processes that sustain them. This knowledge is essential for fields such as medicine, biotechnology, and environmental science, and it provides a foundation for exploring the complexities of life itself.

10.1. Exploring the role of chemistry in everyday life

Chemistry is a central science that plays a fundamental role in our everyday lives. It underpins the natural processes and materials we encounter daily and enables many modern conveniences and technological advancements. From the food we eat to the products we use, chemistry is everywhere, impacting our health, comfort, and quality of life.

10.1.1. Food and Nutrition:

- **Nutrient Composition:** Chemistry helps us understand the composition of the food we consume, including macronutrients such as carbohydrates, proteins, and fats, as well as micronutrients like vitamins and minerals.
- **Preservation and Safety:** Chemical preservatives and additives help extend the shelf life of food and improve safety by preventing spoilage and contamination.
- **Cooking and Flavoring:** The chemical reactions that occur during cooking, such as the Maillard reaction, contribute to the flavor, aroma, and appearance of food. Understanding these reactions allows for better culinary techniques and the development of new recipes.

10.1.2. Medicine and Healthcare:

- **Pharmaceuticals:** Chemistry plays a vital role in the development of drugs and medications that treat illnesses and improve health. Medicinal chemistry focuses on designing and synthesizing compounds with therapeutic properties.
- **Diagnostics:** Chemical reactions are used in medical diagnostics, such as blood tests and other laboratory tests, to

detect and measure substances in the body and diagnose health conditions.

- **Personal Care Products:** Chemistry is essential in the formulation of skincare, haircare, and cosmetic products. These products often contain chemical compounds that provide specific benefits, such as moisturization, cleansing, and sun protection.

10.1.3. Cleaning and Hygiene:

- **Household Cleaning Products:** Chemistry is involved in the development of cleaning agents such as detergents, soaps, and disinfectants. These products help maintain hygiene and cleanliness in homes and workplaces.
- **Water Purification:** Chemistry plays a role in treating and purifying water to make it safe for drinking and other uses. Chemical processes such as chlorination and filtration remove contaminants and pathogens.

10.1.4. Energy and Fuels:

- **Fossil Fuels and Alternatives:** Chemistry is involved in the extraction and refinement of fossil fuels such as oil and natural gas. It also plays a role in the development of alternative energy sources, including biofuels and hydrogen.
- **Batteries and Energy Storage:** Chemistry is critical in the design and production of batteries, which store and release energy for various applications, from portable electronics to electric vehicles.

10.1.5. Materials and Manufacturing:

- **Polymers and Plastics:** Chemistry is key to the synthesis of polymers and plastics, which are used in a wide range of products from packaging materials to construction and electronics.
- **Textiles and Fabrics:** The chemical treatment of textiles, such as dyeing and finishing, enhances the appearance, durability, and functionality of fabrics.

10.1.6. Environmental Protection:

- **Pollution Control:** Chemistry helps develop methods to control and reduce pollution, such as catalytic converters for vehicles and scrubbers for industrial emissions.
- **Waste Management:** Chemistry is used in processes such as recycling and composting, contributing to sustainable waste management practices.

In summary, chemistry is a foundational science that influences many aspects of our daily lives, from the food we eat to the products we use. By understanding the role of chemistry in everyday life, we gain insight into how chemical processes and substances contribute to our well-being and shape our world.

10.2. The science behind common chemical reactions and substances

Chemical reactions and substances play an essential role in various aspects of our daily lives, from cooking and cleaning to medical treatments and energy production. Understanding the science behind common chemical reactions and substances helps us appreciate their functions and benefits and use them safely and effectively.

10.2.1. Types of Chemical Reactions:

- **Combustion:** Combustion, or burning, is a chemical reaction between a fuel and an oxidizing agent (typically oxygen), resulting in the release of energy in the form of heat and light. Combustion reactions are fundamental for heating, transportation, and power generation.
- **Acid-Base Reactions:** Acid-base reactions involve the transfer of hydrogen ions (protons) between an acid and a base. These reactions are common in household cleaning products, as well as in biological systems such as the human body, where pH balance is crucial for health.
- **Redox Reactions:** Redox (reduction-oxidation) reactions involve the transfer of electrons between substances. This type of reaction is essential in processes such as cellular respiration, energy production, and metal corrosion.
- **Precipitation Reactions:** Precipitation occurs when two solutions containing soluble salts are mixed, resulting in the formation of an insoluble solid, or precipitate. These reactions are used in water treatment to remove contaminants.
- **Polymerization:** Polymerization is a chemical process in which monomers (small molecules) join together to form polymers

(long chains of molecules). This reaction is key to the production of plastics, synthetic fibers, and other polymers.

10.2.2. Common Chemical Substances and Their Uses:

- **Water (H_2O):** Water is a universal solvent and is involved in many chemical reactions. It is essential for life and is used in a wide range of applications, including drinking, cleaning, and industrial processes.
- **Sodium Chloride (NaCl):** Commonly known as table salt, sodium chloride is used for seasoning food and as a preservative. It also plays a critical role in regulating fluid balance in the body.
- **Carbon Dioxide (CO_2):** Carbon dioxide is a byproduct of respiration and combustion. It is used in the production of carbonated beverages and as a refrigerant. CO_2 also plays a significant role in the greenhouse effect and climate change.
- **Sodium Bicarbonate ($NaHCO_3$):** Also known as baking soda, sodium bicarbonate is used in baking as a leavening agent, as well as for cleaning, deodorizing, and neutralizing acids.
- **Acetic Acid (CH_3COOH):** Commonly known as vinegar, acetic acid is used in cooking, cleaning, and preserving foods. It is also used in the production of chemicals and pharmaceuticals.
- **Hydrogen Peroxide (H_2O_2):** Hydrogen peroxide is a common disinfectant and bleaching agent. It is used for cleaning wounds, sanitizing surfaces, and whitening teeth.

10.2.3. The Role of Catalysts:

- **Catalysts:** Catalysts are substances that speed up chemical reactions without being consumed in the process. They play a

key role in many industrial and biological processes, including the production of fertilizers, fuels, and pharmaceuticals.
- **Enzymes:** Enzymes are biological catalysts that facilitate reactions in living organisms. They are essential for digestion, metabolism, and other cellular processes.

Understanding the science behind common chemical reactions and substances provides insights into their applications and the principles governing their behavior. This knowledge allows us to make informed choices about how to use these substances effectively and safely in our daily lives.

Chapter 11: The Zoology of Our World

Zoology, the scientific study of animals, encompasses the exploration of animal biology, behavior, evolution, and ecology. It examines the incredible diversity of animal life on Earth, ranging from microscopic invertebrates to complex mammals. Understanding zoology gives us insights into the interconnectedness of species and ecosystems, the adaptations that allow animals to survive and thrive, and the impact of human activity on wildlife and natural habitats.

In this chapter, we will explore key aspects of zoology, beginning with the vast diversity of animal species and their classification into taxonomic groups such as phyla, classes, and families. By understanding how animals are classified, we can appreciate their evolutionary relationships and shared characteristics.

Next, we will delve into animal behavior, investigating how animals interact with their environment and each other. Topics such as communication, mating rituals, and social structures will be covered, providing insight into the complex behaviors exhibited by various species.

We will also examine the ecological roles of animals, highlighting the importance of different species in maintaining healthy ecosystems. This includes discussions on predator-prey relationships, symbiotic partnerships, and the roles of herbivores, carnivores, and omnivores in food webs.

The chapter will touch on the process of evolution, explaining how natural selection drives the diversity of life and shapes the adaptations that animals exhibit to survive in their environments. We will explore examples of convergent and divergent evolution, demonstrating how similar traits can arise in different species and how species can diverge into new forms.

Additionally, we will discuss the challenges faced by wildlife today, including habitat loss, climate change, and human-wildlife conflict. Understanding these challenges is crucial for developing conservation strategies to protect endangered species and preserve biodiversity.

Finally, we will look at the ways in which zoology influences fields such as medicine, agriculture, and conservation biology. For instance, studying animal models helps researchers understand human diseases, while insights into animal behavior inform strategies for wildlife management and conservation.

By understanding the zoology of our world, we gain a deeper appreciation for the richness and complexity of animal life on Earth. This knowledge can inspire us to protect and preserve the diverse species and ecosystems that make up our natural world.

11.1. Understanding the diversity of animal life and its impact on our daily lives

The diversity of animal life on Earth, also known as biodiversity, encompasses the wide variety of species, their genetic variability, and the complex ecosystems in which they exist. Animals play critical roles in natural processes and human society, influencing our daily lives in numerous ways.

11.1.1. The Diversity of Animal Life:

- **Taxonomic Classification:** Animals are classified into various taxonomic groups based on shared characteristics and evolutionary relationships. These groups include phyla, classes, orders, families, genera, and species. Understanding this classification helps us appreciate the relationships among different animal groups and their evolutionary history.
- **Species Richness:** The animal kingdom is incredibly diverse, with millions of known species ranging from simple invertebrates like insects and worms to complex vertebrates such as mammals and birds. Each species contributes uniquely to the ecosystem and plays a specific role in maintaining balance.
- **Adaptations:** Animals have evolved a wide range of adaptations to survive and thrive in different environments. These adaptations include specialized body structures, physiological processes, and behaviors that enable animals to cope with challenges such as predators, changing climates, and competition for resources.

11.1.2. Impact on Our Daily Lives:

- **Ecosystem Services:** Animals play vital roles in providing ecosystem services that benefit humans, such as pollination, seed dispersal, and pest control. For example, bees and other pollinators are essential for the production of many fruits, vegetables, and nuts.
- **Food Sources:** Animals provide important sources of food for humans, including meat, fish, and dairy products. Sustainable management of animal populations is crucial to ensure a stable food supply and preserve biodiversity.
- **Medicine:** Many medical treatments and pharmaceuticals are derived from animals. For example, compounds from animals like venomous snakes and marine organisms are used in drug development. Studying animal biology also helps us understand human diseases and develop treatments.
- **Cultural and Recreational Value:** Animals hold cultural significance in many societies and are a source of inspiration for art, literature, and folklore. Wildlife watching and other recreational activities centered around animals contribute to human well-being and economic development.
- **Companionship:** Domesticated animals such as dogs and cats provide companionship and emotional support for millions of people worldwide, enhancing mental health and quality of life.
- **Scientific Research:** Animals are often used in scientific research as models to study biological processes and develop new technologies. Ethical considerations and regulations guide the use of animals in research to ensure their welfare.
- **Economic Impact:** Animals play a role in industries such as agriculture, fishing, and tourism. For instance, sustainable fisheries provide livelihoods for many people, while ecotourism supports local economies and conservation efforts.

In summary, the diversity of animal life enriches our daily lives in numerous ways, from providing food and medicine to offering cultural and recreational value. Recognizing the importance of animals in our world can inspire us to protect and preserve biodiversity, ensuring the continued health and prosperity of both humans and wildlife.

11.2. The science behind animal behavior and ecosystems

Understanding animal behavior and ecosystems is crucial to gaining insight into how animals interact with one another and their environment, as well as the complex interdependencies within ecosystems. These interactions influence the health and sustainability of natural systems, impacting both wildlife and human society.

11.2.1. Animal Behavior:

Animal behavior encompasses the ways animals interact with their environment, other species, and members of their own species. The study of animal behavior, known as ethology, involves observing and interpreting the actions and reactions of animals in different situations.

- **Communication:** Animals use a variety of methods to communicate with one another, including vocalizations (such as bird calls and whale songs), visual signals (like body language and coloration), chemical signals (pheromones), and tactile signals (touch).
- **Mating and Reproduction:** Mating behaviors vary widely across species, ranging from elaborate courtship rituals (such as dances and displays) to the release of pheromones to attract mates. Reproductive strategies also differ, with some species producing large numbers of offspring, while others invest heavily in parental care.
- **Social Structures:** Many animals live in social groups with complex structures, such as packs, herds, and colonies. Social behavior can include cooperation, competition, and dominance

hierarchies, all of which play a role in survival and reproductive success.

- **Foraging and Feeding:** Animals have evolved various foraging strategies to locate and acquire food efficiently. These strategies may involve hunting, scavenging, or grazing, and often require cooperation and communication within a group.
- **Migration and Navigation:** Many species migrate seasonally to take advantage of changing environmental conditions or food availability. Animals use various cues such as the sun, stars, magnetic fields, and landmarks to navigate long distances.

11.2.2. Ecosystems:

An ecosystem is a community of organisms interacting with their physical environment. These interactions create complex networks of relationships that sustain the ecosystem's balance and function.

- **Food Webs:** Food webs represent the flow of energy and nutrients through an ecosystem, starting with primary producers (plants) and moving through herbivores, carnivores, and decomposers. Each species plays a role in maintaining the stability of the food web.
- **Trophic Levels:** Organisms in an ecosystem can be classified into trophic levels based on their source of energy. Primary producers convert sunlight into energy through photosynthesis, herbivores consume plants, and carnivores eat other animals.
- **Symbiotic Relationships:** Symbiosis refers to close interactions between different species that benefit at least one party. Examples include mutualism (both species benefit), commensalism (one benefits while the other is unaffected), and parasitism (one benefits at the expense of the other).
- **Predator-Prey Dynamics:** Predation is a key ecological interaction that influences the population dynamics of both

predators and prey. These dynamics can have cascading effects throughout the ecosystem.

- **Keystone Species:** Some species have a disproportionate impact on their ecosystem relative to their abundance. Removing or altering the population of a keystone species can lead to significant changes in the structure and function of the entire ecosystem.
- **Human Impact:** Human activities such as habitat destruction, pollution, and climate change can disrupt ecosystems and alter animal behavior. Conservation efforts aim to mitigate these impacts and protect biodiversity.

By studying animal behavior and ecosystems, we gain a deeper understanding of the intricacies of nature and the importance of maintaining ecological balance. This knowledge can inform conservation strategies and sustainable practices that benefit both wildlife and human communities.

Chapter 12: The Botany of Our Surroundings

Botany, the scientific study of plants, provides insights into the life processes, structures, and interactions of the plant kingdom. Plants play a fundamental role in sustaining life on Earth, providing the oxygen we breathe, the food we eat, and the raw materials for many essential products. Understanding the botany of our surroundings helps us appreciate the complexity of plant life and its impact on ecosystems and human society.

In this chapter, we will explore key aspects of botany, beginning with the diversity of plant life and the classification of plants into taxonomic groups such as divisions, classes, orders, families, genera, and species. By understanding plant classification, we can appreciate the evolutionary relationships and shared characteristics of different plant groups.

Next, we will delve into plant physiology and life cycles, examining how plants grow, reproduce, and interact with their environment. Topics such as photosynthesis, respiration, and plant hormones will be covered, providing insight into the essential processes that sustain plant life.

We will also examine the ecological roles of plants, highlighting their importance in maintaining healthy ecosystems. This includes discussions on plants as primary producers in food webs, their role in soil formation and nutrient cycling, and their contributions to biodiversity.

The chapter will touch on the impact of human activity on plants and ecosystems, including habitat destruction, invasive species, and climate change. Understanding these challenges is crucial for developing conservation strategies to protect plant diversity and maintain ecosystem health.

Additionally, we will discuss the ways in which botany influences fields such as agriculture, medicine, and environmental science. For instance, the study of plant genetics and breeding has led to advances in crop production and food security, while plants are also a source of many important medicines.

Finally, we will look at the importance of plants in human culture and well-being, including their aesthetic, recreational, and spiritual value. Plants enrich our lives in countless ways, from providing beautiful landscapes and green spaces to serving as sources of inspiration and relaxation.

By understanding the botany of our surroundings, we gain a deeper appreciation for the vital role plants play in sustaining life on Earth. This knowledge can inspire us to protect and preserve plant diversity and promote sustainable practices that benefit both the environment and human society.

12.1. Exploring the world of plants and their role in our everyday lives

Plants form the foundation of life on Earth, playing essential roles in sustaining ecosystems and supporting human life. From providing oxygen and food to offering natural beauty and medicinal benefits, plants are indispensable to our daily lives.

12.1.1. Plants as Primary Producers:

- **Photosynthesis:** Plants use sunlight, carbon dioxide, and water to produce oxygen and glucose through the process of photosynthesis. This process is the basis of the food chain, providing energy and sustenance to almost all life forms.
- **Oxygen Production:** One of the most crucial contributions of plants is the production of oxygen, which is essential for the survival of aerobic organisms, including humans.

12.1.2. Plants in Food Production:

- **Staple Crops:** Plants provide the majority of our food supply, including staple crops like rice, wheat, and maize, which form the basis of human diets around the world.
- **Fruits and Vegetables:** Plants produce a wide variety of fruits and vegetables that supply essential nutrients, vitamins, and minerals necessary for human health.
- **Herbs and Spices:** Plants provide flavoring agents in the form of herbs and spices, enhancing the taste of food and offering additional health benefits.

12.1.3. Plants in Medicine:

- **Medicinal Plants:** Throughout history, plants have been used as sources of medicine. Many modern pharmaceuticals are derived from plant compounds, such as aspirin (from willow bark) and quinine (from cinchona bark).
- **Herbal Remedies:** Plants are commonly used in traditional medicine practices around the world. Herbal remedies often contain active compounds that offer therapeutic effects.

12.1.4. Plants in Materials and Industry:

- **Building Materials**: Plants provide materials such as wood and bamboo, which are used in construction and furniture making.
- **Fibers:** Plants like cotton, flax, and hemp produce natural fibers used in textiles, ropes, and other products.
- **Dyes and Pigments:** Plants offer natural dyes and pigments used in coloring fabrics, foods, and cosmetics.

12.1.5. Plants in Environmental Protection:

- **Erosion Control:** Plants help stabilize soil and prevent erosion through their root systems.
- **Water Purification:** Wetland plants play a role in filtering pollutants from water, aiding in natural water purification.
- **Climate Regulation:** Plants sequester carbon dioxide from the atmosphere, helping to mitigate climate change and regulate the Earth's climate.

12.1.6. Plants for Aesthetic and Recreational Value:

- **Gardening and Landscaping:** Plants add beauty to our surroundings and improve our quality of life through gardens, parks, and other green spaces.

- **Houseplants:** Indoor plants enhance indoor air quality and contribute to mental well-being.
- **Cultural and Spiritual Significance:** Plants hold cultural and spiritual value in many traditions, used in ceremonies, rituals, and artistic expressions.

In summary, plants play an integral role in our everyday lives, providing essential resources, services, and benefits. By recognizing the importance of plants, we can better appreciate their contributions and work towards conserving plant diversity and promoting sustainable practices for a healthier future.

12.2. The science behind plant growth, photosynthesis, and plant-based products

The science of botany delves into the processes that enable plants to grow, produce food through photosynthesis, and provide plant-based products that sustain human life. Understanding these processes helps us appreciate the essential role plants play in our world.

12.2.1. Plant Growth:

- **Anatomy and Physiology:** Plants have specialized structures such as roots, stems, and leaves that perform various functions necessary for growth. Roots absorb water and nutrients from the soil, stems transport these substances throughout the plant, and leaves capture sunlight for photosynthesis.
- **Plant Hormones:** Plant growth and development are regulated by hormones, including auxins, gibberellins, cytokinins, abscisic acid, and ethylene. These hormones control processes such as cell elongation, seed germination, fruit ripening, and stress responses.
- **Phototropism and Gravitropism:** Plants respond to external stimuli such as light and gravity. Phototropism is the growth of a plant towards light, while gravitropism is the orientation of growth in response to gravity.

12.2.2. Photosynthesis:

- **Process:** Photosynthesis is the process by which plants convert sunlight, carbon dioxide, and water into glucose (a form of sugar) and oxygen. This occurs in the chloroplasts, organelles within plant cells that contain the pigment chlorophyll.

- **Light Reactions:** The light-dependent reactions of photosynthesis occur in the thylakoid membranes of the chloroplasts. These reactions convert solar energy into chemical energy in the form of ATP and NADPH.
- **Calvin Cycle:** The Calvin Cycle, also known as the light-independent reactions, takes place in the stroma of the chloroplasts. In this cycle, carbon dioxide is fixed into organic molecules using the energy from ATP and NADPH to produce glucose.
- **Importance:** Photosynthesis is the foundation of the food chain and the primary source of energy for most life forms on Earth. It also produces oxygen, which is essential for aerobic organisms.

12.2.3. Plant-Based Products:

- **Food:** Plants provide a wide variety of foods, including grains, fruits, vegetables, legumes, and nuts. These plant-based foods supply essential nutrients for human health.
- **Medicinal Compounds:** Many plants contain bioactive compounds with medicinal properties. For example, digitalis (from foxglove) is used to treat heart conditions, and artemisinin (from sweet wormwood) is used to treat malaria.
- **Fibers:** Plants such as cotton, flax, hemp, and jute produce natural fibers used in textiles, ropes, and other products.
- **Wood and Timber:** Trees provide wood and timber used for construction, furniture, paper, and other products.
- **Oils and Resins:** Plants produce oils and resins used in cooking, cosmetics, and industrial applications. Examples include olive oil, palm oil, and pine resin.

- **Dyes and Pigments:** Plants offer natural dyes and pigments used in coloring fabrics, foods, and cosmetics. Examples include indigo, madder, and saffron.

12.2.4. Advancements in Plant Science:

- **Genetic Engineering and Breeding:** Advances in plant genetics and breeding have led to the development of crops with improved yields, disease resistance, and nutritional content.
- **Sustainable Agriculture:** Plant science research focuses on sustainable farming practices such as crop rotation, cover cropping, and precision agriculture to enhance productivity while minimizing environmental impact.

Understanding the science behind plant growth, photosynthesis, and plant-based products allows us to appreciate the critical role plants play in sustaining life on Earth. This knowledge can guide sustainable practices and innovations that benefit both the environment and human society.

Chapter 13: The Mathematics of Finance

The mathematics of finance is a field of study that uses mathematical concepts and models to analyze financial markets, manage risk, and make investment decisions. Understanding the mathematics of finance can help individuals and institutions make informed choices about saving, investing, and planning for the future.

In this chapter, we will explore the key concepts and mathematical tools used in finance, including interest rates, compounding, annuities, and amortization. These concepts form the foundation of financial planning and investment strategies.

Next, we will delve into the analysis of financial markets, examining the mathematical models used to assess risk and return in various types of investments such as stocks, bonds, and derivatives. This includes an exploration of concepts such as expected returns, variance, and standard deviation.

We will also discuss the principles of portfolio management, including diversification and asset allocation. These strategies help investors balance risk and return in their investment portfolios.

The chapter will touch on the time value of money, a key concept in finance that explains how the value of money changes over time. This includes discussions on present value and future value calculations, which are essential for evaluating investments and financial decisions.

Additionally, we will explore the use of mathematical models in risk management, such as options pricing models and value-at-risk (VaR) calculations. These tools help investors and financial institutions manage and mitigate financial risks.

Finally, we will discuss the ethical considerations and challenges in the mathematics of finance, such as the importance of transparency and fairness in financial practices.

By understanding the mathematics of finance, individuals and institutions can make more informed financial decisions and develop effective strategies for managing wealth and risk. This knowledge is essential for navigating the complexities of financial markets and achieving long-term financial goals.

13.1. Understanding financial concepts like interest, investment, and risk through mathematics

Mathematics plays a vital role in understanding financial concepts such as interest, investment, and risk. By applying mathematical principles to finance, individuals and institutions can make informed decisions about managing money, investing, and planning for the future.

13.1.1. *Interest:*

- **Simple Interest:** Simple interest is calculated on the initial principal amount over a specified period. The formula for simple interest is $I = P \times r \times t$, where I is the interest, P is the principal amount, r is the interest rate, and t is the time in years. Simple interest does not consider the effects of compounding.
- **Compound Interest:** Compound interest is calculated on the principal amount and the accumulated interest over time. The formula for compound interest is $A = P \times (1 + r/n)^{(nxt)}$, where A is the future value of the investment, P is the principal amount, r is the annual interest rate, n is the number of compounding periods per year, and t is the time in years. Compound interest can lead to exponential growth of investments over time.
- **Interest Rates:** Interest rates play a crucial role in finance, affecting borrowing costs, savings rates, and investment returns. Central banks use interest rates to influence economic growth and inflation.

13.1.2. *Investment:*

- **Return on Investment (ROI):** ROI measures the profitability of an investment. It is calculated as the percentage gain or loss relative to the initial investment, using the formula ROI = (Net Profit / Cost of Investment) x 100%.
- **Net Present Value (NPV):** NPV evaluates an investment by calculating the present value of future cash flows, minus the initial investment. A positive NPV indicates that an investment is expected to generate more cash flows than the cost.
- **Internal Rate of Return (IRR):** IRR is the discount rate that makes the NPV of an investment equal to zero. It represents the expected annualized return on an investment.

13.1.3. Risk:

- **Risk Assessment:** Financial risk involves the potential for loss in investments. Mathematics is used to measure and analyze risks such as market risk, credit risk, and liquidity risk.
- **Standard Deviation:** Standard deviation is a measure of the dispersion or variability of returns. It is often used as an indicator of risk, with higher standard deviation implying higher risk.
- **Expected Returns:** Expected returns are calculated using probabilities and the potential returns of an investment. This measure helps investors evaluate the potential profitability of an investment.
- **Value at Risk (VaR):** VaR is a statistical measure that estimates the potential loss in value of an investment or portfolio over a specific time period with a given level of confidence. It is commonly used for risk management.

13.1.4. Diversification:

- **Portfolio Diversification:** Diversification involves spreading investments across different asset classes and sectors to reduce risk. Mathematical models such as modern portfolio theory (MPT) are used to optimize asset allocation and achieve the best possible return for a given level of risk.
- **Correlation:** Correlation measures the relationship between the returns of different investments. Diversifying across assets with low or negative correlation can help minimize overall portfolio risk.

By understanding financial concepts like interest, investment, and risk through mathematics, individuals and institutions can make more informed financial decisions. This knowledge is essential for achieving long-term financial goals and managing risk effectively in an ever-changing financial landscape.

13.2. The role of maths in personal finance and economic forecasting

Mathematics plays a central role in personal finance and economic forecasting, enabling individuals and organizations to make informed decisions and plan for the future. By using mathematical tools and models, people can manage their finances effectively, and economists can predict economic trends and fluctuations.

13.2.1. Math in Personal Finance:

- **Budgeting and Expense Tracking:** Mathematics is used to create and manage budgets, tracking income and expenses to ensure that individuals and families live within their means. Simple arithmetic helps balance accounts and make informed decisions about spending and saving.
- **Saving and Investing:** Mathematical calculations are essential for saving and investing. Compounding interest calculations help individuals understand the growth of savings and investments over time.
- **Loan Calculations:** Mathematics is used to calculate monthly loan payments, interest rates, and total cost of loans. Understanding loan terms and amortization schedules helps borrowers make informed decisions and avoid excessive debt.
- **Retirement Planning:** Retirement planning involves projecting future income and expenses to determine how much to save for retirement. Using models that account for inflation, expected returns, and life expectancy, individuals can plan their savings and investments to maintain their desired lifestyle.
- **Tax Planning:** Calculating taxes accurately and planning for tax liabilities are essential aspects of personal finance.

Mathematics helps determine taxable income, deductions, and potential tax savings.

- **Insurance:** Mathematical models such as probability and statistics help assess risks and set insurance premiums. Understanding these concepts can help individuals choose the right insurance policies for their needs.

13.2.2. Math in Economic Forecasting:

- **Statistical Analysis:** Economists use statistical methods to analyze economic data and identify trends. Techniques such as regression analysis help forecast economic variables like GDP growth, inflation, and unemployment rates.
- **Time Series Analysis:** Time series analysis involves examining data over time to identify patterns and make predictions about future trends. This approach is commonly used in economic forecasting for variables such as stock prices, interest rates, and exchange rates.
- **Economic Indicators:** Math is used to analyze and interpret economic indicators such as consumer price index (CPI), producer price index (PPI), and housing starts. These indicators provide insights into the overall health of the economy and guide economic policy decisions.
- **Modeling Economic Relationships:** Economists use mathematical models to understand relationships between different economic variables. For example, models like the Phillips curve explore the trade-off between inflation and unemployment.
- **Risk Assessment and Management:** Mathematics helps assess and manage economic risks such as market volatility and credit risk. Models such as Value at Risk (VaR) and stress testing are used to evaluate the potential impact of economic shocks.

- **Policy Analysis:** Mathematical models are used to evaluate the potential impact of economic policies on variables like inflation, employment, and GDP growth. This analysis helps policymakers make informed decisions.
- **Monte Carlo Simulations:** Monte Carlo simulations involve using random sampling to estimate the probability of various outcomes. Economists use these simulations to model complex systems, such as the potential range of returns on an investment portfolio or the likelihood of different economic scenarios.
- **Decision Trees and Bayesian Analysis:** Decision trees and Bayesian analysis are used in economic forecasting to model complex decision-making processes. These tools help economists and financial analysts understand the impact of different choices and uncertainties on economic outcomes.
- **Machine Learning and AI:** Recent advancements in machine learning and artificial intelligence have revolutionized economic forecasting. These technologies use large datasets and advanced algorithms to identify patterns and make predictions about economic trends.

13.2.3. Implications for Individuals and Policymakers:

- **Informed Financial Decisions:** By understanding and utilizing mathematical concepts in personal finance, individuals can make more informed decisions about budgeting, saving, investing, and planning for retirement.
- **Risk Management:** Math allows individuals and institutions to assess and manage risk in their financial portfolios and business operations, helping them protect their investments and assets.
- **Policy Development:** Economic forecasting provides policymakers with data-driven insights to design and

implement effective policies that promote economic stability and growth.

- **Market Efficiency:** Economic forecasting can lead to more efficient markets by providing valuable information about future trends. This information helps guide investment decisions and improves the allocation of resources.
- **Education and Literacy:** Financial literacy is closely tied to mathematics, as understanding mathematical concepts is essential for making sound financial decisions. Promoting financial literacy can empower individuals to manage their finances more effectively.
- **Global Economic Impacts:** Math-driven economic forecasting and analysis play a crucial role in understanding global economic trends and their potential impacts on local economies. This understanding can guide international cooperation and development efforts.

By understanding the role of math in personal finance and economic forecasting, individuals and organizations can make better financial decisions and plan effectively for the future. Mathematical models and analyses provide the foundation for sound financial management and economic planning.

In conclusion, the role of math in personal finance and economic forecasting is vast and multifaceted. Mathematics provides the foundation for understanding and managing personal finances, as well as for predicting and analyzing economic trends. By leveraging mathematical tools and models, individuals and policymakers can make more informed and strategic decisions, leading to greater financial security and economic stability.

Chapter 14: The Physics of Energy

Energy is a fundamental concept in physics, representing the ability to perform work or produce change. It exists in various forms, such as kinetic energy, potential energy, thermal energy, and chemical energy, and can be transferred or transformed from one form to another. Understanding the physics of energy is essential for comprehending the principles behind natural phenomena, technological advancements, and the functioning of the universe.

In this chapter, we will explore the different forms of energy and how they are measured and quantified. This includes a discussion of the laws of thermodynamics, which describe the conservation and transfer of energy in different systems.

Next, we will examine the various sources of energy, including fossil fuels, nuclear energy, and renewable sources such as solar, wind, and hydroelectric power. We will discuss how these sources are harnessed and converted into usable forms of energy.

We will also delve into the efficiency of energy conversion processes, exploring concepts such as the Carnot cycle and the efficiency limits of different energy systems. Understanding these concepts is crucial for developing sustainable energy technologies.

The chapter will touch on the environmental impact of energy production and consumption, including issues such as greenhouse gas emissions, pollution, and resource depletion. These challenges underscore the importance of transitioning to cleaner and more sustainable energy sources.

Additionally, we will explore the role of energy in various applications, such as transportation, industry, and residential use. This includes a discussion of energy storage and distribution systems, such as batteries and the electrical grid.

Finally, we will discuss the future of energy, including emerging technologies such as fusion power and advanced renewable energy systems. These innovations have the potential to revolutionize the way we produce and consume energy, paving the way for a more sustainable future.

By understanding the physics of energy, we gain insight into the processes that power our world and shape our lives. This knowledge is essential for making informed decisions about energy use, policy, and technology development for a sustainable future.

14.1. Exploring the principles of energy conservation and transformation

Energy conservation and transformation are foundational principles in physics that describe how energy is transferred and converted from one form to another while remaining consistent in a closed system. These principles are essential for understanding natural phenomena and technological processes.

14.1.1. Law of Conservation of Energy:

- **Principle:** The law of conservation of energy, also known as the first law of thermodynamics, states that energy cannot be created or destroyed in an isolated system. It can only be transformed from one form to another.
- **Closed System:** In a closed system, the total energy remains constant over time, although it can change forms. For example, potential energy can be converted into kinetic energy when an object falls due to gravity.
- **Energy Transfers:** Energy can be transferred between different objects or systems through various means such as work and heat transfer. For instance, when a person lifts a weight, they transfer energy from their muscles to the weight, doing work on it and giving it potential energy.

14.1.2. Forms of Energy:

- **Kinetic Energy:** Kinetic energy is the energy an object possesses due to its motion. The formula for kinetic energy is $KE = 1/2\ mv^2$, where m is the object's mass and v is its velocity.
- **Potential Energy:** Potential energy is the energy an object possesses due to its position or configuration. Examples include

gravitational potential energy (due to height) and elastic
potential energy (stored in a stretched spring).

- **Thermal Energy:** Thermal energy, also known as heat energy, is
  the energy of an object due to the motion of its particles. It is
  related to temperature and is often transferred between
  objects of different temperatures.
- **Chemical Energy:** Chemical energy is stored in the bonds of
  chemical compounds. It can be released during chemical
  reactions, such as the combustion of fossil fuels.
- **Nuclear Energy:** Nuclear energy is stored in the nucleus of an
  atom and can be released through nuclear fission (splitting of
  atomic nuclei) or nuclear fusion (combining atomic nuclei).

14.1.3. Energy Transformation:

- **Converting Energy Forms:** Energy can be transformed from one
  form to another in various processes. For example, in a power
  plant, chemical energy from fuel is converted into thermal
  energy, which is then transformed into mechanical energy
  through a turbine, and finally into electrical energy via a
  generator.
- **Efficiency of Energy Conversion:** The efficiency of energy
  conversion processes can vary, with some forms of energy
  conversion being more efficient than others. For example,
  electric motors and transformers can achieve high efficiency,
  while some thermal engines have lower efficiency due to heat
  loss.
- **Carnot Cycle:** The Carnot cycle is an idealized thermodynamic
  cycle that defines the maximum efficiency achievable by a heat
  engine. Real-world engines operate at lower efficiencies due to
  practical limitations.

14.1.4. Energy Conservation in Everyday Life:

- **Reducing Energy Consumption:** Energy conservation in daily life can be achieved by using energy-efficient appliances, insulating homes, and adopting renewable energy sources such as solar panels.
- **Sustainable Practices:** Conserving energy helps reduce environmental impact and resource depletion. By minimizing energy waste and maximizing efficiency, individuals and organizations can contribute to a sustainable future.

Understanding the principles of energy conservation and transformation is key to making informed decisions about energy use and technology development. It allows us to better manage resources, minimize waste, and develop sustainable energy systems for a cleaner and more efficient future.

14.2. The role of physics in understanding renewable energy sources and climate change

Physics plays a crucial role in understanding renewable energy sources and climate change, providing insights into the mechanisms behind energy generation and the impact of human activity on the environment. By applying principles of physics, we can develop sustainable energy technologies and address the challenges of climate change.

14.2.1. Understanding Renewable Energy Sources:

- **Solar Energy:** Physics helps us understand how to harness solar energy using photovoltaic cells (solar panels) and solar thermal systems. Photovoltaic cells convert sunlight directly into electricity, while solar thermal systems use sunlight to heat a fluid, which can then be used to generate electricity or provide heat.
- **Wind Energy:** Wind energy is generated by converting the kinetic energy of moving air into mechanical energy using wind turbines. The design and optimization of wind turbines involve principles of aerodynamics, fluid dynamics, and mechanical engineering.
- **Hydropower:** Hydropower involves converting the potential energy of water flowing from higher to lower elevations into electricity. This process requires an understanding of fluid mechanics and the design of turbines and generators.
- **Geothermal Energy:** Geothermal energy is harnessed from heat stored within the Earth. This process involves drilling wells to access hot water or steam, which can be used for heating or to drive turbines and generate electricity.

- **Biomass Energy:** Biomass energy is produced by converting organic matter, such as plant material and waste, into heat or electricity. This process often involves combustion or biological conversion, such as anaerobic digestion, which can be optimized using principles of thermodynamics and chemistry.

14.2.2. Physics and Climate Change:

- **Greenhouse Effect:** Physics helps us understand the greenhouse effect, which occurs when certain gases in the atmosphere, such as carbon dioxide and methane, trap heat from the Sun. This leads to an increase in global temperatures and contributes to climate change.
- **Climate Modeling:** Climate models use mathematical equations based on the laws of physics to simulate and predict climate behavior over time. These models help scientists understand how human activities and natural processes impact the climate.
- **Energy Balance:** Physics provides insights into the Earth's energy balance, which is the difference between incoming solar radiation and outgoing heat radiation. Changes in this balance, such as increased greenhouse gas emissions, can lead to changes in global temperatures.
- **Sea Level Rise:** Physics helps us understand the processes driving sea level rise, including the melting of glaciers and ice caps and the thermal expansion of seawater. These changes are linked to rising global temperatures.
- **Mitigation Strategies:** Physics-based solutions play a critical role in developing technologies and strategies to mitigate climate change. For example, the use of renewable energy sources reduces the reliance on fossil fuels and lowers carbon emissions.

<u>**14.2.3. *Advancements and Challenges:***</u>

- **Energy Storage:** Efficient energy storage solutions, such as batteries and other technologies, are essential for integrating renewable energy sources into the grid. Physics informs the design and optimization of these storage systems.
- **Grid Integration:** Physics helps optimize the integration of renewable energy sources into the electrical grid, including managing supply and demand, ensuring grid stability, and minimizing losses.
- **Efficiency Improvements:** Physics-based research contributes to improvements in the efficiency of renewable energy technologies, such as more efficient solar cells and wind turbine designs.

In summary, physics provides the foundation for understanding and advancing renewable energy sources while also helping us comprehend and address the causes and consequences of climate change. By leveraging this knowledge, we can work toward a more sustainable energy future and mitigate the impacts of climate change on our planet.

Chapter 15: The IT of the Future

The future of information technology (IT) promises transformative changes that will impact every aspect of our lives. From artificial intelligence and quantum computing to the Internet of Things and blockchain technology, emerging IT innovations will revolutionize how we interact with technology, data, and one another. This chapter explores the key areas that will shape the future of IT and the potential implications of these advancements.

We will begin by examining the evolution of IT infrastructure, including the shift toward cloud computing and edge computing. These paradigms enable faster, more efficient processing and data storage while supporting the growing demand for distributed systems and real-time data analysis.

Next, we will delve into the rapid development of artificial intelligence (AI) and machine learning (ML) technologies. These fields are expected to play a central role in automating complex tasks, improving decision-making processes, and enabling new applications across industries.

The chapter will also explore the rise of quantum computing, which has the potential to revolutionize computing power and solve complex problems that are currently beyond the reach of classical computers. This breakthrough technology could have profound implications for cryptography, drug discovery, and optimization.

The Internet of Things (IoT) is another area of focus, as the proliferation of connected devices creates opportunities for new services and data-driven insights. However, this also raises challenges related to security, privacy, and data management.

Blockchain technology, known for its use in cryptocurrencies like Bitcoin, offers promising applications beyond finance. Its potential to enhance security, transparency, and trust in digital transactions could disrupt a wide range of industries.

Human-computer interaction (HCI) is another key area of development, with emerging technologies such as augmented reality (AR) and virtual reality (VR) changing how we interact with digital content and the physical world.

Lastly, we will discuss the ethical, social, and regulatory challenges associated with the future of IT. Issues such as data privacy, AI ethics, and the digital divide require careful consideration to ensure that technological advancements benefit society as a whole.

By understanding these emerging trends and challenges, we can better prepare for the opportunities and responsibilities that the future of IT holds. This knowledge is essential for shaping a technological future that is innovative, inclusive, and sustainable.

15.1. Predicting future trends in information technology

Predicting future trends in information technology (IT) requires careful analysis of emerging technologies, societal needs, and industry developments. While it is impossible to predict the future with absolute certainty, several key trends are likely to shape the IT landscape in the coming years.

15.1.1. Artificial Intelligence and Machine Learning:

- **Advanced AI Models:** AI and machine learning (ML) are expected to continue evolving, with more advanced models capable of understanding context and performing complex tasks. These models will become increasingly integrated into various industries, from healthcare to finance.
- **AI Ethics and Regulation:** As AI becomes more prevalent, ethical considerations and regulatory frameworks will play a larger role in guiding its development and use. Transparency, fairness, and accountability will be key priorities.
- **AI Augmentation:** AI will increasingly be used to augment human capabilities in decision-making, creativity, and productivity. This trend could lead to significant changes in how work is done across various industries.

15.1.2. Quantum Computing:

- **Increased Research and Development:** Quantum computing research will continue to gain momentum, with advancements in quantum hardware, algorithms, and error correction. These breakthroughs could lead to significant performance improvements.

- **Applications:** As quantum computing matures, it will find applications in areas such as cryptography, materials science, and optimization problems, potentially revolutionizing industries that rely on these technologies.

15.1.3. Internet of Things (IoT):

- **Proliferation of Connected Devices:** The number of IoT devices is expected to grow exponentially, enabling new services and applications in smart homes, smart cities, and industrial automation.
- **Security and Privacy Concerns:** As more devices become connected, security and privacy will become increasingly important. Developing secure IoT ecosystems will be a top priority.
- **Edge Computing:** Edge computing will play a crucial role in managing the vast amounts of data generated by IoT devices, enabling real-time processing and analysis closer to the data source.

15.1.4. Blockchain Technology:

- **Beyond Cryptocurrencies:** Blockchain technology will continue to expand beyond cryptocurrencies, finding applications in supply chain management, healthcare, and identity verification, among others.
- **Interoperability and Scalability:** Developing interoperable and scalable blockchain networks will be crucial for enabling widespread adoption and realizing the full potential of the technology.

15.1.5. Human-Computer Interaction (HCI):

- **Augmented Reality (AR) and Virtual Reality (VR):** AR and VR technologies will become more immersive and integrated into daily life, transforming industries such as gaming, education, and healthcare.
- **Natural User Interfaces:** Voice, gesture, and gaze-based interfaces will become more prevalent, enhancing user experience and making technology more accessible.

15.1.6. Cloud and Edge Computing:

- **Hybrid and Multi-Cloud Solutions:** Organizations will increasingly adopt hybrid and multi-cloud strategies to optimize performance, cost, and flexibility.
- **Serverless Computing:** Serverless architectures, which allow developers to focus on code rather than infrastructure, will continue to gain popularity for their scalability and cost-effectiveness.

15.1.7. Ethical and Regulatory Challenges:

- **Data Privacy:** Ensuring data privacy will remain a major concern, particularly as IT becomes more integrated into daily life.
- **Ethical AI:** Ethical considerations in AI, including bias mitigation and transparency, will be critical to ensure equitable and responsible use of technology.
- **Regulation and Governance:** As IT technologies mature, regulations and governance will be necessary to guide their use responsibly.

Predicting future trends in IT involves keeping an eye on current technological developments, industry trends, and societal needs. By understanding these trends, businesses, researchers, and policymakers can prepare for the future and guide the development of IT in a direction that is beneficial for society as a whole.

15.1.8. Cybersecurity and Privacy:

- **Adaptive Security:** With the growing complexity of cyber threats, adaptive security measures will become essential for protecting systems and data. AI-driven security solutions and threat intelligence will play a key role in identifying and mitigating risks.
- **Zero Trust Architecture:** The adoption of zero trust security models, which require verification for every user and device attempting to access resources, will increase as organizations prioritize security and data privacy.
- **Privacy-Preserving Technologies:** As data privacy becomes a top concern, technologies such as differential privacy, homomorphic encryption, and secure multi-party computation will gain prominence for protecting sensitive information.

15.1.9. Advanced Networking:

- **5G and Beyond:** The widespread adoption of 5G technology will enable faster, more reliable connectivity and support the proliferation of IoT devices and high-bandwidth applications.
- **Software-Defined Networking (SDN):** SDN and network function virtualization (NFV) will continue to transform networking, offering greater flexibility, scalability, and efficiency.

15.1.10. Automation and Robotics:

- **Robotic Process Automation (RPA):** RPA will see increased adoption in businesses for automating repetitive tasks, improving efficiency, and reducing errors.
- **Collaborative Robots (Cobots):** Cobots, designed to work safely alongside humans, will find applications in manufacturing, healthcare, and other industries, enhancing productivity and safety.

15.1.11. Personalized Experiences and Services:

- **Data-Driven Personalization:** IT will enable increasingly personalized experiences and services across various industries, such as healthcare, entertainment, and retail. This will be driven by data analytics and machine learning.
- **Customer-Centric Solutions:** Businesses will focus on creating customer-centric solutions, using technology to understand and meet individual needs more effectively.

15.1.12. Education and Workforce Development:

- **Lifelong Learning:** The IT industry will continue to evolve rapidly, necessitating ongoing education and training for professionals to stay up-to-date with new technologies.
- **Digital Skills Training:** As IT becomes more pervasive, digital literacy and technical skills training will be essential for preparing the workforce for future job roles.

In summary, predicting future trends in information technology involves observing current technological advancements and anticipating how they may evolve and integrate into various aspects of society. While the future holds exciting possibilities for innovation and growth, it is also essential to address challenges such as ethics, privacy, and security to ensure a sustainable and equitable technological future. By staying informed and proactive, individuals and organizations can harness the potential of IT to drive positive change and advancement.

15.2. The potential impact of quantum computing, virtual reality, and other emerging technologies

Emerging technologies such as quantum computing, virtual reality (VR), and others have the potential to dramatically reshape various industries and aspects of our lives. These innovations promise to bring about breakthroughs in computing power, data analysis, and user experiences.

15.2.1. Quantum Computing:

- **Computational Power:** Quantum computing uses the principles of quantum mechanics to perform complex computations that classical computers struggle with. It has the potential to revolutionize fields such as cryptography, materials science, and optimization.
- **Cryptography:** Quantum computing could both pose a threat to existing cryptographic methods and enable new forms of secure communication. Organizations will need to prepare for quantum-resistant encryption.
- **Drug Discovery:** Quantum computing can accelerate the drug discovery process by simulating molecular structures and interactions more efficiently than classical computers.
- **Financial Modeling and Risk Analysis:** The complex algorithms required for financial modeling and risk analysis can benefit from quantum computing's ability to process large datasets quickly and accurately.

15.2.2. Virtual Reality (VR) and Augmented Reality (AR):

- **Immersive Experiences:** VR and AR offer immersive and interactive experiences in gaming, education, training, and entertainment. They allow users to engage with digital environments in new and innovative ways.

- **Healthcare and Therapy:** VR and AR have applications in healthcare, such as medical training, surgery simulations, and therapy for patients with phobias or post-traumatic stress disorder (PTSD).
- **Remote Collaboration and Communication:** VR and AR can transform remote work and communication by providing a more immersive and interactive environment for collaboration and meetings.
- **Retail and Real Estate:** AR can enhance the online shopping experience by allowing users to virtually try on clothes or visualize furniture in their homes before purchasing. In real estate, AR and VR can offer virtual tours of properties.

15.2.3. Other Emerging Technologies:

- **Blockchain Technology:** Beyond cryptocurrencies, blockchain has the potential to revolutionize industries by providing secure, transparent, and immutable records of transactions. This can improve supply chain management, identity verification, and digital rights management.
- **Internet of Things (IoT):** IoT enables the interconnection of everyday objects, creating smart homes, cities, and industries. This can lead to more efficient resource management, better safety, and new opportunities for data-driven insights.
- **Edge Computing:** Edge computing brings data processing closer to the source, reducing latency and enabling real-time analytics. This is crucial for applications such as autonomous vehicles and smart cities.
- **5G Technology:** The rollout of 5G networks will enable faster and more reliable connectivity, driving advancements in IoT,

AR, and VR. It will also support emerging applications such as smart cities and autonomous vehicles.

15.2.4. Challenges and Considerations:

- **Ethical Concerns:** Emerging technologies raise ethical questions about privacy, data security, and the potential for misuse. Responsible development and regulation will be essential.
- **Accessibility:** Ensuring that new technologies are accessible to all segments of society will be important to avoid exacerbating existing inequalities.
- **Workforce Impact:** As emerging technologies disrupt industries, there may be significant impacts on the workforce. Reskilling and upskilling initiatives will be necessary to prepare workers for new roles.
- **Regulation and Governance:** Governments and international bodies will need to establish frameworks to guide the ethical and responsible development and use of emerging technologies.

The potential impact of quantum computing, virtual reality, and other emerging technologies is vast and varied. While these innovations promise significant advancements across industries, they also present challenges that need to be carefully managed to ensure that their benefits are realized responsibly and equitably. The impact of emerging technologies, including quantum computing, virtual reality (VR), and others, is vast and varied. While these innovations promise significant advancements across industries, they also present challenges that need to be carefully managed to ensure that their benefits are realized responsibly and equitably.

15.2.5. AI and Machine Learning:

- **Automation and Productivity:** AI and machine learning (ML) can automate routine tasks, allowing employees to focus on more complex and creative work. This can lead to increased productivity across industries.
- **Personalization and Recommendation Systems:** AI and ML enable more personalized experiences in areas such as e-commerce, streaming services, and digital advertising. These systems can improve user satisfaction but must be designed to respect privacy and avoid bias.
- **AI Ethics:** As AI becomes more widespread, ethical considerations such as transparency, accountability, and fairness will be crucial. Ensuring that AI systems do not perpetuate or amplify existing biases is a significant challenge.

15.2.6. Human-Computer Interaction (HCI) and Natural Interfaces:

- **Voice and Gesture Control:** Natural interfaces such as voice and gesture control are becoming more prevalent in consumer electronics and industrial applications. These interfaces improve accessibility and usability, making technology more intuitive for a wide range of users.
- **Brain-Computer Interfaces (BCI):** BCI technology, while still in its early stages, holds the potential to enable direct communication between the human brain and machines. This could lead to revolutionary applications in healthcare and assistive technology.

15.2.7. Biotechnology and Genomics:

- **Precision Medicine:** Advances in genomics and biotechnology, combined with AI, are paving the way for precision medicine,

which tailors medical treatment to an individual's genetic
makeup. This could improve patient outcomes and reduce side
effects.

- **Bioinformatics:** The use of data analytics and AI in genomics
  and proteomics can accelerate research and drug development,
  leading to new treatments and therapies.
- **Ethical and Privacy Concerns:** The rise of genomics and
  biotechnology brings ethical and privacy concerns related to
  genetic data ownership, use, and potential misuse.

15.2.8. Cybersecurity and Privacy:

- **Evolving Threat Landscape:** Emerging technologies also
  introduce new challenges in cybersecurity, as they create new
  attack surfaces and vulnerabilities. Ensuring robust security
  measures is essential for maintaining trust and safety in digital
  environments.
- **Privacy Protections:** As technology continues to advance,
  ensuring user privacy becomes increasingly important.
  Regulations such as the General Data Protection Regulation
  (GDPR) are crucial for establishing privacy standards and
  protecting individuals' data.

In summary, emerging technologies such as quantum computing, VR,
AI, and others hold immense potential for transforming industries
and daily life. They promise breakthroughs in areas such as
healthcare, education, entertainment, and finance. However, these
technologies also come with challenges related to ethics,
accessibility, workforce impact, and regulation. As society navigates
these changes, it will be important to balance innovation with
responsibility to ensure a future that is inclusive, equitable, and
sustainable.

End Description:

As you reach the end of "The Science We Live By: Understanding Our Everyday World," take a moment to reflect on the journey we've started together. Throughout these pages, we've explored the intricate shades of science that sustains our daily lives, from the mathematics of time to the physics of light, the wonders of information technology, and the mysteries of chemistry, biology, and beyond.

What started as a quest to unravel the mysteries of the world around us has transformed into a profound appreciation for the beauty and complexity of the scientific principles that control our existence. Through the lens of mathematics, physics, and technology, we've peeled back the layers of everyday phenomena, revealing the underlying logic and order that shape our experiences.

But beyond the mere understanding of scientific concepts lies a deeper message — a call to embrace curiosity, to question the world around us, and to approach life with a rational and curious mindset. As an enthusiastic supporter of logic and science I invite you to join me in embracing the wonders of the natural world and the transformative power of knowledge.

So, dear reader, how do you feel as you reach the end of this journey? Have you gained a newfound appreciation for the science we live by, or perhaps discovered a passion for exploration and discovery? Take a moment to reflect on the insights you've gleaned and the questions that have been sparked within you.

As you close the final pages of this book, remember that the journey doesn't end here. Let the curiosity that has been ignited within you serve as a guiding light as you continue to explore the wonders of the world around you. Whether you find yourself questioning the mathematics of nature, wondering at the physics of light, or delving into the mysteries of the universe, know that the pursuit of knowledge is a lifelong adventure — one that promises endless discovery and boundless wonder.

Thank you for accompanying me on this enlightening journey through "The Science We Live By." May the insights you've gained and the questions you've asked continue to inspire you long after you've turned the final page. Until we meet again, keep seeking, keep questioning, and above all, keep exploring the extraordinary world of science that surrounds us every day.

Happy Exploring:)